Artificial Intelligence-Based Infrared Thermal Image Processing and Its Applications

Infrared thermography is a fast and non-invasive technology that provides a map of the temperature distribution on the body's surface. This book provides a description of designing and developing a computer-assisted diagnosis (CAD) system based on thermography for diagnosing such common ailments as rheumatoid arthritis (RA), diabetes complications, and fever. It also introduces applications of machine-learning and deep-learning methods in the development of CAD systems.

Key Features:

- Covers applications of various image processing techniques in thermal imaging applications for the diagnosis of different medical conditions
- Describes the development of a computer diagnostics system (CAD) based on thermographic data
- Discusses deep-learning models for accurate diagnosis of various diseases
- Includes new aspects in rheumatoid arthritis and diabetes research using advanced analytical tools
- Reviews application of feature fusion algorithms and feature reduction algorithms for accurate classification of images

This book is aimed at researchers and graduate students in biomedical engineering, medicine, image processing, and CAD.

Artificial Intelligence-Based Infrared Thermal Image Processing and Its Applications

U. Snekhalatha

K. Palani Thanaraj

Kurt Ammer

CRC Press
Taylor & Francis Group
Boca Raton London New York

CRC Press is an imprint of the
Taylor & Francis Group, an **informa** business

First edition published 2023
by CRC Press
6000 Broken Sound Parkway NW, Suite 300, Boca Raton, FL 33487-2742

and by CRC Press
4 Park Square, Milton Park, Abingdon, Oxon, OX14 4RN

CRC Press is an imprint of Taylor & Francis Group, LLC

© 2023 U. Snekhalatha, K. Palani Thanaraj and Kurt Ammer

ISBN: 978-1-032-15814-3 (hbk)
ISBN: 978-1-032-15817-4 (pbk)
ISBN: 978-1-003-24578-0 (ebk)

DOI: 10.1201/9781003245780

Typeset in Times
by SPi Technologies India Pvt Ltd (Straive)

Contents

Preface.. xi
Acknowledgments... xiii
Authors... xv

Chapter 1   Fundamentals of Infrared Thermal Imaging 1

    1.1   Basics of Thermometry 2
    1.2   Thermal Properties of Biological Tissues.............................. 3
    1.3   Thermal Signals Associated with Physiological
        Functions 4
        1.3.1   Temperature Signal of Respiration.......................... 5
        1.3.2   Thermal Heart Rate Signal 9
        1.3.3   Identification of Sweat... 11
        1.3.4   Facial Muscle Activation Signals 14
    1.4   Examples of Recent Applications of Infrared Thermal
        Imaging in Medicine.. 16
        1.4.1   Rheumatoid Arthritis 16
        1.4.2   Diabetes Mellitus... 18
        1.4.3   Dermatology... 18
        1.4.4   Dental Disorders... 19
        1.4.5   Complex Regional Pain Syndrome 21
        1.4.6   Fever Screening.. 23
    1.5   Summary.. 25
    References 26

Chapter 2   Protocol for Standardized Data Collection in Humans 29

    2.1   Background and International Guidelines 29
    2.2   Thermal Camera Performance Check 30
    2.3   Ambient Temperature Control.. 33
        2.3.1   Choice of Ambient Temperature 38
        2.3.2   Structure of the Examination Space 38
    2.4   Subject Selection 40
        2.4.1   Procedures Performed during the Examination..... 41
        2.4.2   Behavior Prior to the Thermal Imaging................ 42
        2.4.3   Data Checking at the Beginning of the Imaging
            Session... 42
    2.5   Patient Position and Image Acquisition.............................. 43
        2.5.1   Static or Dynamic Thermal Imaging..................... 45
    2.6   Thermal Image Analysis... 47
    References 50

Chapter 3   Basic Approaches of Artificial Intelligence and Machine
                 Learning in Thermal Image Processing...55

        3.1   Image Source 55
        3.2   Data Transfer from Camera to the Analyzing Software 55
        3.3   Pre-processing 56
                 3.3.1   Thermal Image Color Modes 57
                 3.3.2   Image Enhancement...58
                            3.3.2.1   Image Smoothening and Sharpening..... 58
                            3.3.2.2   Histogram Equalization 59
                 3.3.3   Multispectral Images 60
                 3.3.4   Image Registration 60
                 3.3.5   Image Fusion...62
        3.4   Segmentation Techniques 63
                 3.4.1   Clustering-Based Segmentation Algorithms 64
                            3.4.1.1   k-Means Clustering...............................64
                            3.4.1.2   Hierarchical Clustering..........................65
                            3.4.1.3   Divisive Clustering 65
                            3.4.1.4   Density-Based Clustering 66
                            3.4.1.5   Fuzzy C-Means Clustering 66
                            3.4.1.6   Neutrosophic C-Means Clustering 67
                            3.4.1.7   Mean Shift Clustering...........................67
                 3.4.2   Threshold-Based Segmentation............................67
                 3.4.3   Region-Based Segmentation Algorithms 68
                            3.4.3.1   Region Growing...................................69
                            3.4.3.2   Region Merging 69
                 3.4.4   Active Contour 69
                 3.4.5   Edge Detection for Image Segmentation...............70
                            3.4.5.1   Robert Edge Detection..........................70
                            3.4.5.2   Sobel Edge Detection 70
                            3.4.5.3   Prewitt Edge Detection 71
                            3.4.5.4   Marr–Hildreth Edge Detection 72
                            3.4.5.5   Canny Edge Detection 72
                 3.4.6   Watershed Segmentation 73
                 3.4.7   Deep Learning Models for Image Segmentation ... 73
        3.5   Artifact Suppression 73
        3.6   Feature Extraction and Classification...........................74
                 3.6.1   Gray Level Co-occurrence Matrix........................74
                 3.6.2   Speeded-Up Robust Features 75
                 3.6.3   Gray Level Run Length Matrix 76
                 3.6.4   Local Binary Pattern...77
                 3.6.5   Gray Level Size Zone Matrix 77
                 3.6.6   Local Binary Gray Level Co-occurrence Matrix..... 78
                 3.6.7   Scale-Invariant Feature Transform 79
                 3.6.8   Local Directional Pattern.....................................79

       3.6.9   Segmentation-Based Fractal Texture
                Analysis 80
3.7   Machine Learning Classifiers 80
       3.7.1   Logistic Regression 80
       3.7.2   Naïve Bayes Classifier 81
       3.7.3   K-Nearest Neighbors 81
       3.7.4   Decision Tree 81
       3.7.5   Support Vector Machine 82
       3.7.6   Random Forest 82
       3.7.7   Bagging 83
       3.7.8   Logit Boost 83
       3.7.9   K-Star 83
3.8   Deep Learning Classifier 84
       3.8.1   Background 84
       3.8.2   Basic Architecture 85
               3.8.2.1   Convolution layer 85
               3.8.2.2   Rectified Linear Unit (ReLU) 86
               3.8.2.3   Pooling Layer 86
               3.8.2.4   Max Pooling 86
               3.8.2.5   Average Pooling 86
       3.8.3   Batch Normalization 87
       3.8.4   Drop Out 87
       3.8.5   Fully Connected Layer 87
       3.8.6   Training Procedure for CNN 87
       3.8.7   Pre-Processing and Augmentation of Data 87
       3.8.8   Initialization of Parameters 88
               3.8.8.1   Random Initialization 88
               3.8.8.2   Unsupervised Pre-Training
                        Initialization 88
       3.8.9   CNN Regularization 88
       3.8.10   Choosing an Optimizer 89
               3.8.10.1 Gradient Descent 89
               3.8.10.2 AdaDelta 89
               3.8.10.3 Adam 89
3.9   Performance Metrics 89
       3.9.1   Confusion Matrix 89
       3.9.2   Accuracy 90
3.10 Pre-Trained Network Models 91
       3.10.1   VGG Architecture 91
       3.10.2   ResNet V2 92
       3.10.3   DenseNet 121 (Densely Connected
                Convolutional Networks) 92
       3.10.4   MobileNet 93
3.11 Conclusion 93
References 94

Chapter 4   Thermal Imaging for Arthritis Evaluation in a Small
Animal Model 97

  4.1   Induction and Evaluation of Arthritis 97
        4.1.1   Collagen-Induced Arthritis 97
        4.1.2   Mono-Arthritis Model 100
        4.1.3   Pristane-Induced Arthritis 104
        4.1.4   Adjuvant-Induced Arthritis 105
        4.1.5   Antigen-Induced Arthritis 106
        4.1.6   Proteoglycan (PG)-Induced Arthritis 106
        4.1.7   Carrageenin Induced Arthritis 106
  4.2   Thermal Image Acquisition 107
  4.3   Thermal Image Analysis 108
  4.4   Histopathology Evaluation 112
        4.4.1   Histopathological Analysis of Synovial Fluid 112
  4.5   Conclusion 118
  References 119

Chapter 5   Thermal Imaging for Inflammatory Arthritis Evaluation 121

  5.1   Patients and Methods 122
  5.2   Disease Activity Assessment 122
  5.3   Temperature Indices 124
        5.3.1   Heat Distribution Index 125
  5.4   Thermal Image Acquisition Protocol 126
  5.5   Thermal Image Segmentation 130
        5.5.1   Threshold Method 131
        5.5.2   Edge Detection 132
        5.5.3   k-Means Clustering 135
        5.5.4   Watershed Segmentation 136
  5.6   Statistical Feature Extraction and Classification 138
        5.6.1   Oriented Fast and Rotated BRIEF 138
        5.6.2   Support Vector Machine 139
        5.6.3   k-Nearest Neighbor 140
        5.6.4   Linear Discriminant Analysis (LDA) 140
        5.6.5   Random Forest Classifiers 141
        5.6.6   Naïve Bayes Classifier 142
  5.7   Findings and Conclusion 143
  5.8   Case Study 144
        5.8.1   Color Doppler Ultrasonographic Evaluation 146
        5.8.2   Case Reports 147
                5.8.2.1   Case Study 1 147
                5.8.2.2   Case Study 2 148
                5.8.2.3   Case Study 3 149
        5.8.3   Results of CDUS Examination 149

          5.8.4   Comparison of Clinical, Thermographic,
               and CDUS Findings 150
          5.8.5   Discussion 150
          5.8.6   Limitations and Recommendations.................... 151
    References 152

Chapter 6   Potential of Thermal Imaging to Detect Complications in
                Diabetes: Rationale for Diabetes Screening with
                Thermal Imaging 155

    6.1   Introduction 155
          6.1.1   Diagnosis... 156
          6.1.2   Thermography in Diabetes Evaluation............... 158
    6.2   Selection of Study Population 159
          6.2.1   Subject Preparation.. 160
    6.3   Measurement Protocol.. 161
    6.4   Study of Thermograms in Subjects with Diabetes 161
          6.4.1   Facial Thermal Imaging Measurements.............. 162
               6.4.1.1   Temperature Measurement
                           Analysis 162
          6.4.2   Thermal Imaging in Diabetic Foot 166
    6.5   Computer-Aided System for Screening of Diabetic
          Complications.. 173
          6.5.1   Features 174
                6.5.1.1   Feature Extraction and
                           Classification....................................... 174
                6.5.1.2   Oriented Fast and Rotated Brief
                           (ORB) 174
          6.5.2   Automated Deep Learning Classifier 176
                6.5.2.1   Pre-trained CNN Classifier 176
    6.6   Conclusion... 177
    References... 178

Chapter 7   Thermal Imaging in Detection of Fever for Infectious
                Diseases 181

    7.1   Human Body Temperature.. 181
          7.1.1   Fever Development Due to Infectious
                Diseases.. 184
               7.1.1.1   Thermal Imagers for Elevated Body
                           Temperature Detection......................... 184
          7.1.2   Measurement Protocol... 186
    7.2   Fever Detection Systems 191
          7.2.1   Conventional Methods... 191

          7.2.2   AI-Based Fever Screening Systems 192
                   7.2.2.1   Face and Eye Detection 193
      7.3   Conclusion 200
      Acknowledgment 201
      References 202

Chapter 8   Ethical Aspects in Thermal Imaging Research 205

      8.1   Introduction 205
      8.2   General Ethical Principles 207
      8.3   Guidelines for Good Clinical Research Practice 208
      8.4   Responsible Conduct of Research 211
      8.5   Ethical Issues Review Procedure 211
      8.6   Informed Consent Process 212
      8.7   Issues Related to Medical Datasets 212
      8.8   General Guidelines When Using Thermal Imaging for
             Clinical Screening 213
      8.9   Ethical Considerations in Thermal Imaging Research 214
             8.9.1   Infrared Imaging Systems as Medical
                      Devices 214
                      8.9.1.1   Standards of Thermal imaging
                                 system 215
                      8.9.1.2   Handling Thermal Imaging Systems ... 217
             8.9.2   Patient Management Protocol 218
                      8.9.2.1   The Pre-Examination Process 218
             8.9.3   Interpretation and Reporting 219
      8.10   Summary 220
      Appendix I 221
      Appendix II 223
      References 224

Index 225

Preface

Body temperature is a vital parameter used in the medical field for indicating abnormal activity of human tissues. Infrared (IR) thermal imaging produces information on skin temperature distribution in various body regions. The thermal property of human skin varies between the different layers or within the layers of the skin due to the presence of the blood vessels. Depending on the individual's condition and the surrounding environment, the human body will dissipate the excess heat energy. Over the decades, many research works have been carried out to analyze this temperature profile on various skin regions to identify the physiological process behind it. This procedure forms the basis of thermal imaging, whereby advanced thermal imaging sensors are used to capture the heat profile of the human body regions so as to identify the ailments in the human body. With the advancement in thermal imaging systems, high-resolution thermograms could be produced with a temperature sensitivity of less than 50 mK and with a spatial accuracy of around 10 μm. Widespread research is going into developing computer-aided disease diagnosis systems based on IR thermograms as an assisting tool for physicians in clinical screening and evaluation process. The main aim of this book is to provide a detailed description of designing and developing a computer-assisted diagnosis (CAD) system based on thermography for diagnosing some of the common ailments such as rheumatoid aArthritis (RA), diabetes complications, and fever. This book also introduces the readers to applications of machine-learning and deep-learning methods in the development of the CAD system.

This book is organized as follows:

- Chapter 1 deals with the fundamentals of infrared thermal imaging, which includes thermal imaging properties, image acquisition, and various applications.
- Chapter 2 focuses on the standardized protocol to be followed in the selection of test subjects and data acquisition based on infrared thermographs.
- Chapter 3 provides information on image analysis with a focus on image analysis tools and various image processing steps involved in thermal imaging.
- Chapter 4 evaluates arthritis disease progression in the CFA-induced Wistar rat model and monitors inflammatory arthritis activity using thermal imaging compared with histopathological study.
- Chapter 5 showcases the applications of thermal imaging in the diagnosis of rheumatoid arthritis in humans.
- Chapter 6 explains the recent developments in applications of thermal imaging in screening complications in diabetic subjects and discusses their rationales.

- Chapter 7 explains the development of an advanced thermal imaging system that uses deep learning approaches to detect elevated body temperatures for proper fever screening.
- Chapter 8 describes the ethical considerations and data protection in thermal imaging research.

Acknowledgments

First and foremost, I express our heartfelt and deep sense of gratitude to Chancellor Shri. T.R. Pachamuthu, Vice-Chancellor Dr. C. Muthamizhchelvan, Pro-Vice-Chancellor (Admin) Dr. Ravi Pachamoothoo, and Pro-Vice-Chancellor (Academics) Dr. P. Sathyanarayanan of SRM Institute of Science and Technology for providing us the necessary facilities for the completion of our book. I also acknowledge Registrar Dr. S. Ponnusamy of SRM Institute of Science and Technology for his constant support and endorsement.

I wish to express our sincere gratitude to our Dean (Engineering & Technology, Kattankulathur) Dr. T.V. Gopal, and Chairperson (School of Bioengineering) Dr. M. Vairamani of SRM Institute of Science and Technology for their constant support and encouragement.

I am extremely grateful to the Head of the Department, Dr. Varshini Karthik, Biomedical Engineering Department, SRM Institute of Science and Technology, for her invaluable guidance, motivation, and timely and insightful technical discussions. We are immensely grateful for her constant encouragement, smooth approach throughout our journey and to make this work possible. Also, I would like to thank researcher scholars Ms. Richa Rashmi, Ms. Ahalya, Ms. Nithya Kalyani, and my students Ms. Bhargavi and Ms. Sowmiya of Biomedical Engineering Department, SRM Institute of Science and Technology, for their help and suggestions.

I wish to extend my sincere thanks to Dr. J. Kumar, Department of General Medicine, and Dr. Panchapakesa Rajendran, Head, Rheumatology Department, SRM Hospital and Research Center, for providing support and encouragement in collecting the data at SRM Medical College Hospital and Research Center.

I would to thank my family – my husband M. Ravishankar and his parents, my parents, my brothers, and my children – for their continuous love, support, encouragement, and understanding. I thank the Almighty for his blessings.

U. Snekhalatha
SRM Institute of Science and Technology

I would want to take this opportunity to show my gratitude to everyone who has helped us finish this book; to our All-Powerful God for providing us with the strength, knowledge, patience, and wisdom necessary to complete this book.

I would like to express my strong heartfelt gratitude to the Management of St. Joseph's College of Engineering, Chennai, India, for their constant support and encouragement,

I would also want to express our heartfelt gratitude to our colleagues, friends, research scholars, and instructors who have been kind enough to provide recommendations and comments for the text's improvement. Additionally, we would like to express my gratitude to the individuals who provided support and presence, as well as ideas for the development of this book.

K. Palani Thanaraj
St. Joseph's College of Engineering

I like to thank Dr. U. Snekhalatha and Dr. K. Palani Thanaraj for inviting me to contribute to this book. It was a pleasure to work with them, and their open mind for improving this book is much appreciated. I am also grateful for their support and help in editing and correcting the text during a difficult personal health condition.

Prof. Dr. med. Kurt Ammer, PhD
Editor in Chief, *Thermology International*
Treasurer, European Association of Thermology

Authors

Dr. U. Snekhalatha is currently working as an Professor in the Department of Biomedical Engineering, SRM Institute of Science and Technology (SRMIST), Kattankulathur, India. She pursued her Doctorate in Biomedical Engineering at SRMIST (2015). Her areas of interest include biomedical signal processing, medical image processing, biomedical instrumentation, and machine learning and deep learning techniques. She has published 88 research articles in reputed peer-reviewed international journals and international conferences. She has filed four Indian patents of which, all are in published stage. She obtained the Best Researcher Award for her publications in *Nature* indexed journal during the Research Day function held on March 1, 2021 at SRMIST. She received Gandhian Young Technological Innovation Award for paper titled "Design of Acetone Breath Gas Analyzer in the Evaluation of Diabetes Mellitus," Appreciation 2020, Society for Research and Initiatives for Sustainable Technologies and Institutions, November 5, 2020. She has obtained best paper award and gold medals for some research paper publications. She is a life member of various professional societies such as the Biomedical Society of India, IEI, IEANG, IRED, ISTE, and ISCA. She is currently serving as a reviewer and guest editor for various reputed peer-reviewed international journals.

Dr. K. Palani Thanaraj is currently working as an Assistant Professor in the Department of Electronics & Instrumentation Engineering, St. Joseph's College of Engineering, Chennai. He has completed his PhD in 2018 in the Faculty of Information and Communication Engineering from Anna University, Chennai. His research areas include image processing, advanced signal processing, image segmentation, machine learning, and deep learning. He has developed deep learning algorithms for performing image classification of medical images for disease diagnosis. He has published his works in many reputed and refereed journals indexed in Web of Science and Scopus.

Prof Dr. med. Kurt Ammer was certified as a general medical practitioner in 1978, a consultant for physical medicine and rehabilitation in 1989, and a consultant for physical medicine and rehabilitation (rheumatology) in 1994. He was senior researcher at the Ludwig Boltzmann Research Unit for Physical Diagnostics, Austria, between 1988 and 2004. From 1985 until his retirement in early 2013, he was Vice Director of the Institute of Physical Medicine and Rehabilitation at the Hanusch Hospital in Vienna, Austria. He got involved in medical thermography in 1988, and was appointed as secretary and treasurer of the European Association of Thermology (EAT) in 1990, and currently serves as the EAT treasurer. Since 2002, he has been appointed as external professor at the Medical Imaging Research Unit, University of South Wales, Pontypridd, UK. His research interests focus on rehabilitation medicine and the application and standardization of thermal imaging in medicine.

1 Fundamentals of Infrared Thermal Imaging

Infrared thermography is a non-invasive imaging modality which is used to determine the skin surface temperature of any object, since every object emits thermal radiation when it is at a temperature above absolute zero. The heat content of inanimate matter and living beings differs, especially if these living beings have developed a physiological thermoregulation system leading to two different temperature compartments – a central one with maintained temperature and a peripheral shell with constantly changing temperature reflecting the heat exchange with the environment. It is essential to remember that temperature is one of the seven base quantities of the International System of Units (SI). Like pressure and density, temperature is an intensive quantity. A physical property of a system that does not depend on the system size or the amount of material in the system is called intensive. Temperature can simply be seen as the "degree of hotness" or, more specifically the "potential for heat transfer" (Machin and Tsai, 2010).

Like in other objects, heat transfer within and from the human body occurs without a change of state involving conduction and convection. On the surface, additional heat transfer is achieved by thermal radiation or by fluid evaporation; the latter is combined with a change of state. Since the temperature is non-homogeneously distributed across tissues and blood vessels, there is constant heat exchange between regions of different temperatures by conduction and convection. The tissue thermal resistance and bloodstream directions are the important factors that determine the speed of heat transfer. Bloodstream and heat flow may be in opposite directions, resulting in an additional "counterstream" mechanism of heat exchange, uncontrolled by the thermoregulation system. The exchange of heat from skin to air is caused mainly by means of infrared radiation. The radiation emitted by the skin can be captured by a thermal imaging camera and displayed as radiance intensity equal to temperature. Since the anatomical details obtained from the thermal imaging of the human body express the outline structure of the body, it is a common hypothesis that the physiological function in health and disease are reflected by the temperature distribution on the skin.

Several research studies indicated that detection of diseases and disorders could be possible by analyzing the thermal images of subcutaneous skin temperature. Thermal imaging has been used for detecting various medical conditions affecting the locomotor and neuromuscular system such as rheumatoid arthritis and other inflammatory arthritis, osteoarthritis, pain syndromes including

DOI: 10.1201/9781003245780-1

complex regional pain syndrome, orofacial pain, and diabetes complications. Historically, the first medical application of thermal imaging was in breast cancer detection. Among the early fields of thermographic research were also peripheral vascular disorders such as Raynaud's phenomenon or venous diseases, dermatological disorders such as psoriasis or melanoma, and diseases of the locomotor apparatus. Consecutively, the use of infrared thermography was extended to eye diseases, psychological conditions, cardiovascular disorders, metabolic disorders, and, more recently, to physical exercise assessment and mass fever screening.

1.1 BASICS OF THERMOMETRY

Thermometry may be defined as both the branch of physics dealing with the measurement of temperature and the science of the construction and use of thermometers. Physical perception of temperature is the basal experience related to heat exposure. Phenomena such as dilation of air and vapor as a response to heat were already known in ancient Greece, opening the opportunity to detect temperature independent from the human temperature sense (Biró, 2011). Historically, detection and measurement of temperature are based on the laws of classical thermodynamics. In particular, the zeroth law of thermodynamics describes the conditions of thermal equilibrium, in which temperature is the parameter of a state that has the same value for any systems which are in thermal equilibrium. This property is used when the temperature of an object is measured by placing a thermometer in contact with the object, initiating according to the second law of thermodynamics heat flow from high to low temperature that finally results in thermal equilibrium (Machin and Tsai, 2010). Biró (2011) described the ideal relationship between the heat capacities of both the object to be measured and the thermometer to achieve fast thermal equilibration. Other thermometers apply the thermoelectric effects detected by Seebeck, Peltier, and Thompson as an indication of temperature values.

Thermometry can be classified based on the process of heat transfer involved (conductive, convective, or radiative), the field of application (basic science, applied science, industry, biomedicine, biology), or the temperature bands measured. Particularly, radiation thermometers are classified as high-, mid-, or low-temperature measurement devices. Infrared thermal imagers are regarded as an extension of low-temperature radiation thermometers (Zhang and Machin, 2010). Each field of application has standards and limitations that vary between science disciplines. For example, the ISO standard regulating the use and calibration of clinical thermometers is different from the guideline controlling infrared cameras for fever screening. Zhang and Machin described a schematic setup for radiometric temperature measurement. A source of infrared emission is the target for which the temperature is to be measured. An optical system captures the input signal infrared radiation for an infrared-sensitive detector which in turn produces an output signal. The radiation-capturing system is constructed from a lens confining the field of view, a bandpass optical filter determining the spectral region within which the thermal radiation is to be measured, and an aperture in front of the detector to avoid detection of stray light. The output signal (voltage or current) from the detector is amplified and measured by a signal processor. This output signal is related to the spectral radiance emitted by the surface and hence its temperature can be deduced.

Clinical infrared radiometers were first developed in Germany in the 1920s (Cobet, 1926). Most quantitative results of infrared-based clinical temperature measurements in the early phase of infrared thermography were obtained using spot thermometers. Using an optical–mechanical scanning system, thermal cameras at that time were able to generate a monochrome image which was displayed on a cathode ray screen. Since radiometric data were unavailable, infrared thermography was in those days exclusively an imaging modality, and vague ideas existed to analyze thermal images in a quantitative way.

1.2   THERMAL PROPERTIES OF BIOLOGICAL TISSUES

The thermophysical properties of both inanimate material and biological tissues define the conditions for their conductive heat transfer. The properties include thermal conductance defined as the quantity of heat that passes in unit time through a plate of a particular area and thickness when its opposite faces differ in temperature by one kelvin. For a plate of thermal conductivity k, area A, and thickness L, the conductance is kA/L, measured in $W \cdot K^{-1}$. Thermal conductivity k, also symbolized as λ or κ, is a material property, which may be derived from thermal conductance. Thermal resistance R_k is the reciprocal of thermal conductance, L/kA, measured in $K \cdot W^{-1}$. The ratio k/L is called conductive heat transfer coefficient, h_k, measured in $W \cdot K^{-1} \cdot m^{-2}$. The reciprocal of the heat transfer coefficient is thermal insulance, a measurable quantity of importance for the calculation of heat loss.

"Heat capacity" c is a physical property of matter, defined as the amount of heat to be supplied to an object to produce a unit change in its temperature. The SI unit of heat capacity is joule per kelvin (J/K). Thus, heat capacity describes the ability of matter to store thermal energy. Specific heat capacity c_p is found by dividing the heat capacity of an object by its mass. The subscript p in c points to the condition of constant pressure, in which heat supplied to the system would contribute to both the work done and the change in internal energy. Density ρ is defined by the mass of a substance per unit volume. The product of density ρ and specific heat capacity c_p is called volumetric heat capacity $\rho\, c_p$. Finally, thermal diffusivity α combines thermal conductivity with density ρ and specific heat c_p:

$\alpha = \dfrac{k}{\rho c_p}$. At interfaces of contact, the product $k\,\rho\,c$ represents an important feature of the capacities for heat transfer in transient state. The square root of the product of the material's thermal conductivity and volumetric heat capacity is called thermal inertia. The greater the thermal inertia, the higher the heat transfer. Thermal inertia explains why the sensation of the temperature of a given object depends not only on its actual temperature, but even more on its $k\,\rho\,c$ values (Houdas and Ring, 1982).

These physical properties can be used to calculate the passive heat transfer from deep tissues to the surface of a living being. This is a process of physics and not a control mechanism of physiology. It is important not to forget that the aim of thermoregulation is to maintain deep body temperature within narrow regulation. This is achieved by two strategies: changing the thermal resistance of the skin and adaptation of metabolic heat production. Under constant blood pressure, blood flow defined by the ratio of blood volume to tissue volume is dependent on the width of the vascular beds, thus, vasoconstriction reduces blood flow and increases thermal resistance and vasodilation acts vice versa. An additional mechanism for heat dissipation is the evaporation of sweat at the skin, a physiological function available in some, but not all homeotherms. In humans, heat production via cold-induced shivering is well developed, but reduced heat production such as in hibernating mammals is unknown. The skin is an organ that acts as a boundary and interface to the ambiance, but its temperature is not regulated. However, the skin is the place where thermoregulatory responses take place and the site where the heat transfer from deep body structures finally ends. Therefore, if any pathologic process is heat-producing, this thermal energy will be dissipated within the body following the laws of physics, and a small amount of this heat may appear at the skin depending on the location and the intensity of the heat source. Since thermoregulatory responses may interfere with temperature associated with disease processes, it is of advantage to avoid any thermoregulatory responses during capturing medical infrared images for diagnostic purposes.

1.3   THERMAL SIGNALS ASSOCIATED WITH PHYSIOLOGICAL FUNCTIONS

At the skin, heat transfer to the environment is predominately through radiation, which can be captured by an infrared camera and displayed as the temperature on the site from where thermal energy is emitted. The resulting "static" thermal image is analyzed by comparing temperature values obtained in the region(s) of interest (ROI) with a reference temperature, which is often located in the corresponding contralateral anatomical region. However, this temperature signal oscillates in both the time domain and the spatial domain. Michael Anbar was one of the first who subjected these temperature fluctuations using fast Fourier transform (FFT) and relate the detected frequencies to other periodic phenomena such as pulse or neuronal control of blood flow in the microvasculature (Anbar, 1990, 2013). Anbar also showed that the temporal change in surface temperature varies in different anatomical regions.

Although the signal associated with physiological functions is the surface temperature, its distribution is understood as an image feature, and not as a measurable quantity. Consequently, the models applied for extracting physiological functions from surface temperatures are based on image processing.

Respiration contributes to the total heat exchange between the body and the environment by the evaporation of humid air and by convective heat loss through the breathing cycle. Insensible perspiration is the process of a small, but continuous evaporative heat loss via the skin and the respiratory tract. At the skin, this loss is due to a diffusion of water vapor, which can easily be detected by weighing the body (Houdas and Ring, 1982). The predominant source of water content in the skin is sweat production, mainly produced as a thermoregulatory response, but also elicited by various pathologies leading to generalized, regional, or focal hyperhidrosis (Ammer and Ring, 2019). While the water content of the skin due to insensible perspiration provides the basic condition of skin conductivity for direct (galvanic) current, psychogenic sweating contributes to the variation of galvanic skin response (GSR). The periodic blood pressure waves of the arterial pulse are associated with skin temperature alterations. The temperature associated with perfusion is best seen at re-perfusion when external occlusion is released. Autonomic control of microcirculation may be detected in thermal images of the face. This approach attracted recently large interest in psychophysiology (Ioannou et al., 2014), particularly, for the evaluation of emotions (Clay-Warner & Robinson, 2015).

1.3.1 TEMPERATURE SIGNAL OF RESPIRATION

In general, inspired air is cooler and dryer than expired air. Air conditioning occurs mainly in the upper airways and heat exchange is achieved in the small bronchi, and air temperature equals the blood temperature of 37°C in the alveoli. At the nostrils, the temperature of expired air is slightly cooler than it was in the lungs. Thus, the nostrils are a convenient site to obtain the temperature of both inspired and expired air. In conventional respiratory measurements, a thermistor is placed near the nostrils to measure the thermal breathing signal. In thermal imaging, the breathing rate is estimated from the thermal signal detected in ROI located at the nostrils.

Most of the commercially available thermistors are not capable of measuring the tidal volume and find difficulties in tracking the breathing rate accurately over a wide range of breathing frequencies. But the study conducted by Lewis et al. focused on an alternative method to evaluate and monitor the respiration rate and relative tidal volume based on video recordings of facial thermal images. In their study, an automated facial tracking model was used to measure the temperature variations within and around the nostrils. Their obtained thermal signal is converted to quantify the duration and relative depth of each breath cycle.

Lewis et al. (2011) developed a facial tracking algorithm to compensate for head movements and estimated the dominant breathing rate by means of spectral analysis of raw thermal signals obtained in infrared thermal videos of nostrils from 19 study participants performing various breathing patterns. Dominant breathing rate was the basis to calculate the breath-to-breath timing intervals and relative tidal volume. Simultaneously to thermal imaging, respiratory inductance plethysmography was used to estimate respiration rate and tidal volume. Statistically significant correlations were found between the thermal signal and the plethysmograph measures such as breathing interval and relative tidal volume.

The video tracking algorithms are used for automated extraction of thermal signals and to compensate for heat movement. The tracking system has the advantage of performing thermal video analysis, functioning in real time, and identifying the nose features in 3D. Tao and Huang et al. developed the tracking algorithm which is based on the piecewise Bezier volume deformation model (PBVD).

These PBVD algorithms could track the movement of particular facial features. The PBVD model is comprised of a stack of 3D Bezier volumes embedded with the face mesh model. The Bezier volumes are represented by two-layer, non-parallel polygons placed above and below the face mesh model. Image Pyramids were constructed on each video frame and motion parameters were estimated. By means of a tracking algorithm, the thermal signals were extracted from the thermal video. The first frame from the thermal video is displayed and superimposition of 2D coordinates from all vertices in the 3D PBVD model is performed as blue dots over the facial thermal image. Then, the pointer tool is used for the selection of vertex points that enclose the nostrils defined as the ROI. Consequently, an updated 3D model is generated for each video frame with a vertex containing the selected points as a convex polygon.

The tidal volume signal was continuously recorded by using the life shirt (R) inductive plethysmography. Thermal respiratory video signals were recorded for three different breathing methods such as spontaneous, slow and deep, rapid, and shallow. The respiratory signals were extracted from the thermal video signals using the three-step filtering process. The dominant breathing rate was estimated from the recording by means of spectral analysis. Initially, the extended length of 1024 samples was obtained by applying detrending and zero-padding operation. Then, FFT is applied to measure the spectral density distribution. Due to the wide range of breathing frequencies from 3 to 65 cycles/min, the frequency with the highest spectral density was identified as the breathing rate.

As a second step, the thermal signal is converted into a time series signal which is proportional to lung volume. This is achieved by defining the peak of the inspiration signal as the cold temperature and that of the expiration signal as the warm temperature. Thus, by integration of thermal time series, transformed time series is formed which contains the components linearly related to tidal volume. The integrated thermal time series signal was pre-processed using a dynamic filtering process. To accomplish this process, a cubic-polynomial filter with a cut-off frequency less than the breathing rate is considered. The filter cut-off frequency is obtained by using the formula

$$fc = \alpha^* \, \text{BR} \qquad (1.1)$$

where fc is the cut-off frequency, $\alpha = 0.385$, and BR is the breathing rate of the signal.

Fei and Pavlidis (2010) measured the breathing rate in thermal video using automated tracking of the nasal region. They used coalitional tracking algorithm for tracking the facial region during the breathing rate measurements. This can be accomplished by using particle filter trackers which track the dynamic nature of thermal imaging.

In a thermal image of a participant's face, a coalition grid comprising 4 particle filter trackers is placed. The grid outline is marked on the nose region of the facial thermogram by using the click-and-drag method. It is considered as the ROI initialization of the coalitional tracker.

The breathing rate is measured in the nostril region in thermal imaging. During the inspiration process, the subject inhales the cool air from the outside environment. This leads to the convection of heat away from the nostril at an increased rate. Thus, the nostril radiates the heat at a low level toward the camera. During the expiration process, the subject exhales the warm air which is forcefully ejected toward the environment. At this point, heat convection is weak due to the minimal temperature difference observed between the expelled air and the nostril-expired air. Thus, the nostril area radiates the heat at a high level toward the camera. The authors segmented the nostril region from the facial thermograms. The horizontal and vertical edges of the nostril region are detected by applying the Sobel operator. Then, spatial–temporal projections of the nostril in 3D space are obtained for horizontal and vertical projections. Further, they computed the mean horizontal and vertical integral projections for all the frames using the time window for both expiration and inspiration phases. They determined the average temperature within the measurement ROI in every frame. The wavelet transform is applied to the thermal signal to obtain the breathing rate of the signal. Initially, the thermal signal is sampled at a rate of $\delta = 10$ fps. The normalization is applied over the resampled thermal signals to normalize the signal amplitude. Then, continuous wavelet transformation is performed on the normalized thermal signal by using the sliding window. The breathing signal is extracted from the continuous wavelet transform (CWT) analysis. Based on the wavelet scales, the breathing waveform is computed from the nostrils. The breathing rate is estimated by using the center frequency Fc of the mother wavelet and sampling factor $\delta = 10$ fps.

$$\text{Breathing rate} = Fc\,\delta \qquad (1.2)$$

Pavlidis measured the breathing rate of the respiratory signal from the expired air. Another drawback of their method is that raw thermal data were used to build up the signal for each pixel of desired ROI, which affects the signal fidelity due to extreme noise. Hence, to overcome these disadvantages, Chekmenev et al. (2005) indicates the thermal image at different scales and select the important scale which carries relevant information with respect to breathing and heart signal. The authors manually selected the carotid area of the neck as ROI-1 and the nasal area as ROI-2. Then, they decomposed each frame of the thermal images into three average value scales. They calculated the average value of ROI-1 and ROI-2 for each scale of all the frames and obtained 1D plots for the average temperature value.

Further, they applied continuous wavelet analysis on 1D plots.

The CWT is defined as

$$w(a,b) = 1/\sqrt{C\varphi}\;\; 1/\sqrt{a} \int\limits_{-\infty}^{\infty} \varphi * \left(\frac{t-b}{a} \right) f(t)\,dt$$

$$C\varphi = \int\limits_{-\infty}^{\infty} \frac{\varphi^\wedge(\omega)2}{\omega}\,d\omega < \infty \qquad (1.3)$$

$$\varphi^{\wedge}(\omega) = \int_{-\infty}^{\infty} \varphi(t) e^{-j\omega t} \, dt$$

$\varphi(t)$ is mother wavelet.

$$\varphi_{a,t}(t) = \varphi\left(\frac{t-b}{a}\right)$$

where t indicates time, which can also be frame, and b is the time shift.

The reconstructed signal is obtained by using inverse continuous transform.

The average value of the signal is added to the inverse wavelet transform to obtain the reconstructed signal. Based on the prior knowledge of the important structure that belongs to the appropriate scale, the inverse continuous wavelet transform (ICWT) analysis can be performed accurately. Finally, the reconstructed signal can be obtained by adding the average value of the signal to the ICWT. The heart rate is computed by counting the maxima in the reconstructed signal.

Jakkaw and Onoye (2020) monitored the respiratory activity and body movements during the sleep using the thermal imaging technique. The breathing patterns are detected during sleeping and identifying the sleep disorders such as sleep apnea caused due to hypertension, cardiovascular diseases, and arrhythmia. Hence, the authors demonstrated the non-contact method of respiratory and body movement detection using a thermal camera which detects the temperature changes due to breathing patterns. The thermal video frames of subjects in sleeping mode are considered the input images. The images are pre-processed using the Gaussian filter. In the respiration-monitoring method, ROIs are automatically detected from the input images by means of identifying the highest temperature point and massive portions of the high-temperature area. For automated ROI detection of the thermal image, sleeping position is considered. The highest temperature points are detected in the image by using minimum and maximum intensities found in the image. The maximum pixel intensities are associated with the highest temperature of the body. After determining the pixel at the center of the observation area, a rectangular ROI of pixel size either 10×10, 25×25, or 50×50 is applied. Among the ROI used, empirical research result shows that 50×50 pixels produced very good accuracy in compliance with an original frame 640×480. The massive portions of the high-temperature area are detected using the thresholding method. A threshold value of 176 was set, which produced best results in segregating the human skin area from the background.

To extract the respiratory signals, the average of pixel values within each ROI of every frame was calculated from the ROI_h and the ROI_t. The breathing motion is detected by means of finding the difference between the current frame and the past frame or previous frame. Then, the portion of the moved area is extracted using thresholding and morphological operations such as erosion and dilation. Bounding boxes are obtained by identifying the contours after filtering out the noise. The breathing motion is counted based on the number of bounding boxes. The average respiratory signal can be obtained as the combination of root-mean-square calculation of three signals attained from the ROI_h, ROI_t, and breathing motion.

$$\text{Respiratory signal} = \sqrt{ROI_h^2 + ROI_t^2 + BM^2} \qquad (1.4)$$

Then, these fused signals are passed to the Butterworth band pass filter with cut-off frequencies of 0.05 and 1.5 Hz. The filtered signal is further smoothened by using Savitzky–Golay (SG) filter. The moving average is applied to select only the desired peaks. Then, the number of peaks counted depends on the number of breaths.

1.3.2 THERMAL HEART RATE SIGNAL

When the left ventricle pumps the blood from the heart to the other parts of the body, a cardiac pulse is generated. The propagation of cardiac pulse involves mechanical processes as the blood travels through the arterial network and turns back to the heart via the vein network. Hence, Garbey et al. (2007) monitored the pulse waveform by means of skin temperature modulation. This pulsative blood flow modulated the skin tissue temperature due to heat exchange that occurs between the vessels to surrounding tissue by means of convection and conduction. Therefore, the authors proposed a model to simulate the heart diffusion process on the skin's superficial blood vessels and found that the skin temperature waveform is analogous to the pulse waveform. They extracted the cardiac pulse from the external carotid artery or the temporal arterio-venous complex or radial arterio-venous complex. Thus, the modulation of temperature with respect to pulsating blood flow generates the predominant thermal signal on a blood vessel. But this signal is being affected by physiological and environmental factors. Hence, the thermal signal captured by the thermal camera is a composite signal with the pulse considered as one of its components.

In the pulse measurement method, as an initial step, a rectangular ROI is placed over the blood vessel. Temperature profiles of the exposed blood vessel of the subject are obtained. Then, the pixel temperatures along the horizontal direction are averaged. This decreases the noise and constricts the rectangular ROI into a line in which signal measurement initiated. Along the pixels on the measurement line, time evolution signal of its temperature is obtained. Then, FFT is applied to these temperature profiles to attain the power spectra. Then, power spectra of all the profiles are averaged to yield the composite power spectrum.

Kim et al. (2018) proposed a non-invasive heart rate detection based on thermal imaging in the blood vessels of the carotid artery. The blood vessel temperature varies with respect to the blood flow. Hence, it is necessary to identify the blood vessel using a suitable algorithm. The authors used the binarization method to detect the presence of blood vessels. After cropping the required ROI in the thermal image, binarization is applied which produces the output in the form of two-level intensities such as black and white. The background is represented in black color and the foreground object is indicated in white color. Then, the mid-point of the brightest region is measured to identify the position of blood vessels.

Then, they acquire the 1D time series signal at the vessel positions and remove the noise in the signal using bit shift operation toward least significant bit (LSB) detection. Hence, to detect the heart rate in real time, the time series signal is converted into a frequency domain single using discrete Fourier transform (DFT). The heart rate is calculated by choosing the frequency band of 0.9–1.6 Hz corresponding to the heart rate of 54–96 beats/min. The final heart rate is obtained by averaging the multiple heart rate from the multiple points on the identified blood vessels. They compared the heart rate measured from PPG with that of changes in temperature and found that blood vessel is identified correctly only when the estimated accuracy is 95.48%. The accuracy of 78.65% was only achieved when the blood vessels are not identified properly.

Pereira et al. (2018) estimated the heart rate from the thermal videos of mechanical head movements co-occur with the cardiac cycle. As an initial step, pre-processing operations are performed on the thermal images. A multi-level thresholding algorithm was used to segment the head region from the background. According to the algorithm, optimum threshold values are calculated based on the discriminant analysis. The background elimination is followed by contrast enhancement is carried out for all the frames of video thermograms. The next step involves ROI selection particularly focused on the lower part of the head in each frame of thermal video based on the semi-automated method. Then, feature points are chosen from the desired ROI. The authors used Shi–Tomasi corner detector to identify the optimal feature points within the ROI. According to the corner detector, the feature is considered as corner only when the scoring function R is obtained based on the minimum eigenvalues λ_1 and λ_2 greater than the threshold.

$$R = \min\left(\lambda_1, \lambda_2\right) \tag{1.5}$$

$$R > \alpha$$

Initially, the feature points are identified in the first frame of the video sequence. Then, the features are ranked according to the R-score and only the strongest features were selected for tracking. The selected features were given as the input to the template-based point tracker (Kannade Lucas Tomasi tracker) to position the feature point exactly to the head movements. Hence, this tracker identifies all the feature points from the first to the last frame of the video thermal image. As the head movement occurs in the vertical direction, the vertical trajectories of each single feature point over time were computed. After tracking the feature points, temporal filtering is applied to detect vertical trajectories. In that, the low-frequency components dominate the trajectories of feature points due to respiratory movements, whereas high-frequency components provide accurate information for detecting the peak of the heart rate signal. Apart from the vertical movement of the head, other physiological sources such as vestibular activity, facial expression changes, and involuntary muscle movements contribute to the feature point trajectories. Hence, to isolate the heart-related component of interest, it is necessary to decompose other signals into sub-signals. Therefore, the researcher introduced principle component analysis (PCA) to identify the important dimensions with respect to the changes in the head position. They attained this by converting the original dataset into a new data set of uncorrelated variables called principle components. According to the PCA, an appropriate eigen vector is selected to extract the pulse signal. The eigen vectors are ordered with respect to their variance. The first principle component consists of the highest percentage of data variation and i^{th} principle component corresponds to i^{th} highest amount of variation. Hence, the first six principle components are considered for further analysis in identifying the ballistocardiographic head movement. Then, FFT is applied to obtain the amplitude spectra of principle components. As the principle components are chosen based on the periodicity of the signal, the pulse signal is extracted from the frequency components.

1.3.3 IDENTIFICATION OF SWEAT

Since evaporating sweat results in a local reduction of skin temperature, sweat droplets appear in infrared thermal images as cold dots (Ring, 1984). In psychogenic sweating, sweat gland activation may persist in maxillary, nose tip, and perioral regions.

Pavlidis et al. extracted the transpiration signal from thermal images in the perinasal region (Pavlidis et al. 2012) and used the obtained signal as the primary indicator of stress. They quantified the stress based on the estimation of transient perspiratory responses using thermal imaging. The perspiration signal is obtained using a virtual tissue tracker which tracks the ROI even with a few movements. The tissue tracker has the tendency to operate on various head poses and thermal variations. A tracking algorithm is used for identifying the upper orbicularis oris portion of the perinasal region in the initial frame. Spatial–temporal smoothing technique was used in every frame of the thermal clip to estimate the best matching block. To compute the perspiration signal, a morphology-based algorithm was used in each ROI. FFT was used to suppress the high-frequency noise which occurs in the perspiration signal due to imperfections in tracking the signal and breathing effect.

As we know the perspiration signal produced from the perinasal region is the primary indicator of stress, the activated perspiration pores appear as cold dark spots in facial thermal imaging. The morphological operation used to display the cold dark spot is black top-hat transformation. To avoid poor localization of perspiration spots, a contour-based structuring element is used. This contour-based top hat transformation is used to capture the perspiration spots in every frame in the thermal imaging. The instantaneous energy is calculated based on the time at which the frame is captured and the number of detected cold spots at that time. The evolution of instantaneous energy in the perinasal region produced an energy signal which is indicative of perspiration activity.

Eccrine sweat glands are almost widely spread across the whole body. It secretes watery fluid composed of water, NaCl and other solutes onto the skin surface during the emotion stimuli. Evaporation of watery fluid results in abolish of excess heat from the skin surface. Hence, thermal imaging technology is considered as a potential tool in evaluating the variation in cutaneous temperature and its associated emotional states. The regions which are mostly produced a better thermal response to emotional stimuli are the forehead, nose, palm, and maxillary areas. Biological mechanisms such as emotional sweating and subcutaneous vasoconstriction prevent blood flow to the skin surface. This occurs due to different emotions such as pain, fear, and stress, and causes a reduction in temperature in various ROIs. Besides other diagnostic modalities such as GSR, a thermal camera non-invasively picturizing the activity of the sweat gland in the presence of cold dots over the skin surface.

Koroteeva and Bashkatov (2021) were involved in studying the thermal signature related to the excretion of sweat droplet from the sweat pore and its evaporation on the skin surface. Along with the thermal imaging, EEG and GSR were measured using the multi-channel encephalography system. They produced a predominant sweat response from the study participants by means of applying various external stimuli such as sudden mild electric shock, arithmetic mental task, or loud sound to trigger emotional sweating. The thermal videos captured from the facial regions are pre-processed and extracted using FLIR Research IR Max software. As a first step, thermal videos are converted into frames and in each frame, the ROI is fixed on the facial region containing the most predominant active sweat pores. Each ROI contains a single sweat pore. Similarly, a number of ROIs are positioned in every thermal frame. As the second step, an automated algorithm is used to compute the sweat droplet progression using Matlab software. The variation in the sweat droplet with time was measured based on the total number of pixels below a certain threshold and validated with the manual method on the data set.

Further, the authors performed histogram analysis on the highlighted forehead region of the face. They observed that arithmetic exercise provided as mental stimuli to the participants produces significant changes in the sweat gland activity. Due to this, they detected the thermal profiles along the centerline of a single sweat droplet. Similarly, they studied the activity of sweat glands in fingertips with the onset of sudden loud sounds as stimulus. Based on the thermal imaging of the fingertip, the diameter of the sweat droplet on the fingers is less than 1 mm. This size is proportionate to the distance between the two neighboring papillary lines. They observed that the palmar sweat gland has less secretion capacity than the face droplets. They found minimal temperature difference (0.7 K) between the sweat pore and surrounding skin in both face and fingers. Finally, they concluded that emotional sweating due to functioning sweat glands on the face and fingers varies from person to person. Hence, the thermal mapping of active cutaneous sweat glands for each individual person is unique and can be used for biometric identification.

Krzywicki et al. (2014) developed a method for the identification of active sweat glands and monitoring pore activity based on high spatial resolution thermal imaging on finger and face regions. They demonstrated good association between the skin conductance and activity of sweat glands obtained in thermal imaging of eccrine glands in the forehead region. They quantified the pore activity with high spatial resolution in the palm surface of medial, distal, middle, and index fingers of the hand. This approach requires two infrared cameras equipped with lenses of different focal lengths, and a specially designed finger-rest plate. The pore activation index was calculated by means of counting the number of active pores present in the skin area using thermal imaging. They implemented an automated algorithm for counting the active pores using matched filtering techniques and compared it with the manual method. From the thermal image of the active pore, a pore template (7 × 7) is generated. Instead of applying smaller templates of size 5 × 5 or 3 × 3, a 7 × 7 pixel template was identified as the trade-off between the computational efficiency and accuracy of the system. To identify the active pores in the thermal image, the surrounding outer edges of 4 ROIs such as index finger, medial–index finger, medial-middle finger and distal–middle finger were defined manually. The normalized cross-correlation was applied between the pore template and thermal image. If correlation values above a threshold of 0.6 are reached, the PAI is recorded from each thermal video frame. Similarly, PAI was obtained from the facial thermograms. A customized mask was manually designed for each subject. The correlation-based tracking algorithm was used to monitor the skin area. Template matching techniques are used to find the pixel closer to the center of the pore than its edges. If the neighboring pixels near the pore center exceed the correlation values above the threshold, sweat beads are measured.

The automated algorithm is evaluated and compared with the manual method based on the pore count and pore activity on the thumb imaged by thermal video segment by two independent evaluators Thermal videos of active pore regions were captured for 30 s duration by two independent evaluators. Then, the thermal videos are sampled at 25 Hz (750 frames) Considering the sluggish nature of pore responses, the segments were down-sampled to 75 frames to achieve an efficient result at the frame rate of 2.5 Hz. The activation response pores are clearly detectable from the segmented frame of the thermal video. To ensure the accuracy of the test, thermal video segment was independently evaluated by two different evaluators. It was evident that the evaluators are well versed with the qualitative thermal pattern amalgamated with pores of thermal imaging.

The thermal image of middle and index fingers of subject after respiratory work outs indicates the identification of thermal characteristics of the pores. A single frame of each participant helps in providing the representative pore data. The center location of each frame is highly considered as a reference to identify the thermal characteristics of each pore. To identify the center location of a pore, the geometric center of a pore was obtained from the pore activation index. Then, a 7 × 7 window was placed over the geometric center. To fix the exact center point, the temperature and correlation values calculated from the correlation map were attained for each pixel within the 7 × 7 window. The temperature value present in the window were transposed and rescaled so that minimum temperature mapped to "1" and maximum temperature mapped to "0." The correlation coefficients for each pixel in 7 × 7 window are multiplied with the newly mapped temperature to have the combined pixel value of correlation and temperature. From this, the highest pixel value was identified as the center pore location. The authors found a moderate relationship between all PAI measures on the plantar side of the middle and index fingers and skin conductance measured from index and middle fingers of the opposite hand. However, after excluding the results of all low skin conductance responders and considering only last three PAI measurements for correlations, a strong correlation between PAI and SC was obtained. Compared to facial regions, thermal images of fingers provided better information upon sweat gland activation.

Giacinto Di et al. (2014) investigated fear conditioning in post-traumatic stress disorder (PTSD) using a standard GSR method and IR thermal imaging. The authors obtained the gray scale digital images of different emotions such as anger, happiness, and neutral. The participants were requested to count the number of triangles and circles that were projected as images in front of the screen with sudden stimuli, like white noise burst, applied to them. The thermal video images were captured using digital thermal camera FLIR SC 3000 FLIR Systems, Sweden. Then, the sampling process was carried out by setting 10 frames/s. The facial images of the participants were examined qualitatively for their changes in autonomic responses by means of visual inspection. Then, the quantitative analysis of temperature changes in the cutaneous region of the facial thermal image was evaluated using Matlab software. The circular ROI is fixed in the nose region of the thermal image and its corresponding thermal profiles were obtained. The displacement between the ROI locations in successive thermal image frames is corrected by using motion correction and image registration techniques to avoid mismatch in fixing of ROIs. The image registration process involves re-alignment of ROI by using automated recognition and identification of anatomical regions of the nose. After the acoustic stimulation, a decrease in temperature was observed in the entire face due to sudomotor response. The temperature pattern correspondence to emotional sweating response shows the colder dotted spots in nose tip, maxillary, and forehead regions. They found that skin conductance response and facial thermal response provided good investigation methods in the psychometric analysis of PTSD conditions.

1.3.4 FACIAL MUSCLE ACTIVATION SIGNALS

Muscle activation in the face may be indicated by thermal imaging. Human facial expressions result in the activation of muscles which may be associated with changes in skin temperature distribution. The facial action coding system (FACS) and facial surface EMG are the two different methods used to study the characteristics of facial muscle activations. FACS disintegrates facial expressions into discrete action units and combines them together to form more elaborate expressions.

Jarlier et al. (2011) conducted the experiment to study the specific facial muscle heat patterns for examining the sensitivity of thermal imaging. Their ultimate aim is to identify the facial region and perform face normalization followed by feature extraction and facial expression classification based on facial features. During the image acquisition process, the participant's heads were focused toward the imaging system in immobilized condition with a head fixation system. The participants were informed to perform five action units with different intensity–speed combinations. Initially, the acquired images were resized to the facial area alone by using the bilinear interpolation method. The images are aligned with each other using 2D translation and rotation procedures. The images are realigned using an optimization algorithm to obtain the transformation that restores the original image. The affine transformation method is used to match the control point marked in the individual face with the average facial image present in the database. Then, the authors used spatial patterning approach to detect the changes in temperature patterns. They applied spatial PCA to determine the relationship between the temperature values measured for each pixel in the facial regions. Spatial PCA was performed in each trial for participants, considering the pixel as a variable and temperature value as observation. The PCA finds the complex relationship between the temperature values measured for each pixel. The result obtained from PCA is a set of factor loading which corresponds to the spatial component of the original temperature values. The resultant PCA consists of a set of factor loadings from which the first two components were taken into consideration and averaged to produce the facial areas where temperature values depend on action unit performance.

After finding the principle components, the peak temperature (apex) from the facial region is determined. To measure the maximum temperature variations, an automated baseline to peak analysis is performed in order to find the apex value. The speed of the temperature variations has been determined from the latency of the apex. The intensity of temperature variations was obtained from the signal amplitude at the apex. The features extracted from the thermal profiles are a representative temperature map and a mean representative topographic map. The representative temperature map was obtained by subtraction of the temperature value measured at the apex with the baseline temperature. The average representative topographic map is calculated for each action unit by averaging all the individual maps attained in each condition. The quantification of thermal imaging parameters related to facial muscle contractions was classified using the k-nearest neighbor model. The representative temperature map parameter and average representative topographic parameter were given as the input to the classifier to classify the different emotions based on the thermal pattern of facial muscle activity.

Wesley et al. (2012) performed automated facial recognition by comparing the video of thermal imaging, digital facial videos, and fusion of thermal and digital facial images. In their study, the subjects were trained by using a FACS encoder playing the videos of facial expressions and provided sufficient time for practicing each facial expression. Initially, neutral and relaxed expression is captured in both thermal camera and digital camera. Then, the participants were informed to repeat each facial expression 14 times and recorded to simulate various natural expressions. The authors used local feature extraction method in which only the facial areas which are subject to change with facial expressions are considered for processing. To extract the features, ROIs are fixed at 13 fiducial points on the facial region such as forehead, forehead bruise, below the eyes, left and right chin, and cheek. The mean temperature value was computed in each ROI in neural state for the first 25 s. Then, the principle components were extracted from each ROI by considering each pixel value within the ROI as variable. The highest principle

component values are obtained from the frames associated with the largest variation from the neutral ROI. After the determination of principle components from
all ROIs, a profile for each facial expression is obtained by finding the standard
deviation of each ROI. These facial expressions indicate the deviation over the
course of expression. This standard deviation obtained from the profile was used
for training the feed-forward multi-layer perceptrons for both thermal and visual
images. A decision-level fusion method is used for the fusion of thermal and
visual facial images. From their results, it was observed that visual facial images
produced better performance only under the constant lighting conditions, but thermal imaging modality produced good results even if there is heat variation in the
dataset. However, the fusion of thermal and visual imaging modalities has produced better performance compared to individual imaging modalities.

1.4 EXAMPLES OF RECENT APPLICATIONS OF INFRARED
THERMAL IMAGING IN MEDICINE

The following section summarizes articles reporting thermographic studies
employing modern infrared imagers and artificial intelligence methods to detect
joint inflammation, skin diseases, dental or metabolic disorders, or as a method
for fever screening. The camera specifications employed in each application are
collected in a table, and the applied method of image processing is also described.
Finally, the clinical value of each study is critically discussed with respect to the
performance of thermal imaging as a diagnostic tool or as an outcome measure.

1.4.1 RHEUMATOID ARTHRITIS

Infrared thermal imaging has been used for more than five decades for the assessment of inflammatory disorders in various small and large joints of the human
body. In 1975, Francis Ring reported in his M.Sc. thesis the fundamentals for
applying the technique in rheumatology research and practice (Ring, 1975).
Michael Engel published in 1984, reference values for unaffected and inflamed
joints including the wrist, metacarpophalangeal joints, proximal interphalangeal
joints, elbow, knees, ankles, metatarsophalangeal joints, and sternoclavicular and
ileosacral joints (Engel and Saier, 1984). The early thermographic studies in rheumatology suffer from the fact that thermal findings were seldom related to clinical
signs of inflammatory arthritis, although first disease activity scores such as the
Lansbury index were already available.

Pauk et al. (2019a) investigated the hands of RA patients with high and moderate disease activity in comparison to healthy controls using infrared thermal camera FLIR E60bx with a sensitivity of 0.405°C and an accuracy of less than 2% and
thermal resolution of 0.03 K. The authors recorded the thermograms in the finger
region in three different stages: static temperature, thermal video recordings during cooling, and recording during the rewarming process. The authors performed
image pre-processing operations such as normalization and filtering. Then, they
applied morphological erosion and dilation to extract the finger boundaries. Later,
a skeletonization algorithm and a modified depth-first search algorithm were

applied to extract the required ROI and to remove the outer layers of the hand image. The resulting ROI is the one-pixel wide middle line in each finger. They obtained different temperature findings between the three groups investigated. However, the study design missed detecting classical signs of joint inflammation such as swelling (tumor), redness (rubor), and warmth (calor), which can easily be captured by inspection and palpation. In addition, extending the linear ROI to the fingertips beyond the distal interphalangeal joint, which is not affected by rheumatoid arthritis, may have resulted in the overestimation of finger temperature in healthy subjects, who present typically with a fingertip temperature $\pm 0.5°C$ of the metacarpophalangeal temperature. The difference between maximum and minimum temperatures obtained in the finger (Pauk et al. 2019b) was used to compress the temperature information of the linear ROI into a single value but it cannot localize the site of maximal temperature.

It is essential for automatic generation of ROI that they are placed at a meaningful anatomical region. In case of rheumatoid arthritis hands, the recommended positions are the wrist, and the metacarpophalangeal and proximal interphalangeal joints 1 (thumb) to 5 (little finger). Such an array of ROI would allow investigating the coincidence of hyperthermic ROIs with tender or swollen joints, since this information is already available in the individual records of the number and distribution of affected joints. Although time-consuming, manual delineation of ROIs may manage this task better than the automatic generation of measurement areas. The number of hyperthermic joints might become an alternative to the number of swollen or tender joints for the evaluation of disease activity. Unfortunately, data supporting this approach are yet not available.

Snekhalatha et al. (2015, 2018) evaluated the RA using thermal imaging of the hand region. They used hand-held thermal camera ThermaCAM T400 with a resolution of 320×420 and a temperature sensitivity of <50 mK at $30°C$. They implemented k-means and fuzzy c-means algorithm for the segmentation of hand region in RA and normal subjects. They extracted the statistical features using gray level co-occurrence matrix algorithm. The authors developed computer-aided diagnostic tool for automated classification of images from RA patients or healthy subjects. They predicted that the performance of k-means algorithm was better compared to fuzzy c-means in the segmentation of hand thermal images for the total studied population. They found a significant correlation between the skin temperature measured at 2nd and 3rd MCP finger joints with features extracted from segmented hand regions. The main limitation of this study is the lack of coincidence of temperature increase beyond a defined threshold with clinical features such as joint swelling or tenderness. The Second and third MCP are frequently affected by RA and the correlation between extracted features and temperature at this site was found to be significant, however, the agreement of these findings with clinical signs must be questioned. Particularly, the example of the thermal image of a rheumatoid arthritis with homogeneously distributed temperatures between 35.4 and $35.8°C$ on the hand from the interdigital skinfold up to the distal forearm is not a typical finding in rheumatoid arthritis where a focal hyperthermia above the inflamed joints would be expected.

Gatt et al. (2020) studied the thermal characteristics of RA feet during the clinical remission stage. The authors acquired the thermal images in the foot region using a FLIR T630 camera. The thermal camera has a focal plane array with a resolution of 640 × 480 pixels and a sensitivity of <40 mK. They obtained the mean temperatures in the following ROI such as medial, lateral, and central feet and heel regions. The authors found a statistically significant difference between RA and normal subjects in all the forefeet regions. Like in hands, however, focal hyperthermia is only expected in skin regions in close vicinity to inflamed joints, projecting to the distal medial and lateral foot region, and the observed increase in foot sole temperature may not be necessarily caused by the pathology involved in rheumatoid arthritis. Gatt et al. concluded that large, multi-center randomized clinical trials are needed in order to determine the sensitivity and specificity of this imaging modality to finally establish whether it is actually valid and reliable, sufficient to be applied to clinical practice as a screening tool for the timely detection of joint inflammations.

1.4.2 DIABETES MELLITUS

Gururajaroa et al. (2019) conducted a study for the detection of foot ulcers in diabetic neuropathy patients, which describes the conditions for developing software to assess the foot temperatures and differentiates the feet of healthy subjects from the that of ones with diabetes. The authors applied active contour deformable model in combination with morphological processing for the segmentation of foot of patients with diabetes. They extracted the ROI from the segmented image and used the bounding box method to partition the left and right foot regions. They extracted the features and performed asymmetry analysis to compare the left and right foot regions. They also implemented a pre-trained CNN called Mobile net for the classification of normal, diabetes with, and without complications. However, clinical features of healthy controls, diabetes, and diabetes complications are not reported. The significance of the detected variation in temperature of the plantar foot between the investigated three groups remains unknown since the investigated population was small despite data augmentation.

Liu et al. (2015) developed an imaging system composed of thermal and visual color images that is capable to perform an automatic analysis of the plantar feet of diabetes patients. They used a FLIR SC305 thermal camera with a resolution 320 × 240 pixels and an accuracy of ± 2°C. The authors performed foot segmentation in color images using k-means clustering algorithm. They also performed an image registration process to align the segmented color image with the foot thermal image. Based on asymmetry analysis, the system missed only 2 out of 37 foot ulcers in the thermal images.

1.4.3 DERMATOLOGY

Cellulite is an alteration of the topography of the skin that occurs mainly in women in the pelvic region, lower limbs, and abdomen. It is characterized by a padded or "orange peel" appearance (Rossi and Vergnanini, 2000). Bauer et al. (2018) conducted a feasibility study on the utility of infrared thermography for the classification of cellulite stages. They used thermal camera FLIR T335 which has a resolution of 320 × 240 pixels and a sensitivity of 50 mK. They automatically adjusted all recorded thermal images to the same temperature window. They found that cellulite regions exhibit non-uniform temperature distribution in thigh and lower back regions characterized by contours and shapes with uneven and higher temperatures than the surroundings. Those irregularities were marked manually using Image J software. They extracted four characteristics from the thermal images, such as Group 1: Number of irregularities; Group 2: Cumulative area; Group3: Cumulative area of irregularities/area of thigh region; Group4: Irregularities number × area of irregularities. These features were used in a deep learning algorithm to differentiate cellulite stages, which were after further modification of the classifiers able to detect the grade of cellulite correctly in 48 out of 49 thermal images. Unfortunately, scoring the clinical assessment of cellulite patients is incompletely reported, particularly the number of false-positive and false-negative cases of the thermographic evaluation remains obscured. Nevertheless, this promising imaging-orientated approach warrants further investigations.

Herman and Cetingul (2011) acquired images of pigmented lesions with both digital and thermal cameras using a square adhesive marker for the identification of the lesion's location. A landmark detection algorithm was used to identify the corners of the square marker. Further image processing included a random walker segmentation algorithm to create a mask image for the identification of skin lesions and a quadratic motion model to compensate motion-induced artifacts. Mild convective cooling was initiated after the baseline images have been captured. The temperature recovery was observed, and different rewarming patterns were reported for healthy skin and lesioned skin presenting with signs of malignancy in the histological evaluation of tissue obtained from biopsy.

1.4.4 DENTAL DISORDERS

Ammoush et al. (2018) conducted the study to evaluate the role of thermography in the differentiation of patients with suspected dental abscess or facial cellulitis. The authors used a FLIR C2 infrared camera to acquire the thermal image of both lateral views of the face. They obtained higher temperature differences of 2.4°C in facial cellulitis patients compared to dental abscess patients whose temperature difference was 1.5°C between the normal site and the affected site. The absence of criteria for diagnosing cellulitis is the main weakness of this study, but the association of increased temperature in the skin overlying suspected odontogenic inflammation warrants further investigation. Macianskyte et al. (2019) examined the relationship between infrared thermography and computed tomography results by comparing asymmetrical temperature distribution obtained from infrared thermal imaging with the CT lesions. Initially, they detected the human face edges using image gradients and convolution max. They identified the facial symmetry axis to separate the left and right facial regions and calculated the average mean temperature on both the sides of facial regions. They performed image segmentation of left and right ROI using the binary thresholding method based on neighboring temperatures. They obtained a higher average temperature difference in the facial region compared to the mouth region in tumor patients. They observed a negligible temperature difference in the facial region than the mouth region in healthy subjects. The authors predicted that the asymmetrical temperature zone detected from thermal images matches or aligns with the presence of maxilla-facial pathologies obtained from CT images. However, some maximal temperatures at the lesion site were 0.8–2.0°C higher than the core temperature of 37.0°C, questioning the accuracy of temperature readings extracted from the thermal images. In addition, the evaluation of the diagnostic accuracy of temperature asymmetry for the detection of tumorous lesions visible in CT images is unclear in terms of thresholds for both temperatures and spatial distribution. These findings are in contradiction to both the thermodynamics and thermal physiology of heat transfer to the surface.

Kaspryzk et al. (2021) observed the temperature course in buccal skin before, immediately, 1 and 7 days after surgical extraction of a retained third molar. They carried out the thermal imaging with the head positioned in the sagittal plane using thermal camera FLIR E60. They defined two circular ROIs, a large one including the total bucca, and a small one at the projection site of the third molar. A decrease in temperature of 1.5°C was observed from day 1 to day 4 and a sudden increase in temperature occurred from day 4 to day 7. Unfortunately, the authors did not report the findings of clinical inspection of the wound. Thus, their interpretation of skin temperature changes is completely based on the typical histological findings during wound healing. Relating the initial drop in temperature to "repair processes" and the following temperature increase to granular tissue formation is not yet confirmed by controlled studies. Finally, the authors expressed their opinion that thermal imaging might be helpful in monitoring patients after tooth extraction.

Iosif et al. (2016) investigated candida-associated denture stomatitis using infrared thermography based on the comparison of normal subjects and a diseased group. The authors used a ThermaCAM PM 350 infrared camera to acquire thermal images from the maxillary denture bearing area. The authors found statistically significant higher temperatures in candida-associated denture stomatitis (36.20°C) compared to the healthy oral mucosa (34.85°C). The authors reported an optimal thermal threshold of 35.4°C for differentiating normal oral mucosa from the mucosa in candida-associated denture stomatitis. Finally, they concluded that infrared thermography can be used as a complementary investigative method to diagnose inflammatory mouth disorders such as candida-associated denture stomatitis.

Haddad et al. (2016) proposed reference values for normal temperatures at specific sites of the face. The authors used ThermaCAM T400 FLIR systems to acquire the facial thermal image in frontal, lateral, and medial views. They captured thermal images with a size of 10x10 wide window and identified the hottest areas on the forehead, around the eyes and the nose, and in the vicinity of the mouth. Then, they compressed the temperature window by 50% and selected the unchanged hot areas as the site of temperature reference. A strange definition of thermal gradient was used as rationale for selecting the measurement sites. Since a thermal gradient is defined by the temperature difference along a distance, the mean or maximum temperature obtained from a small circular measurement area does not describe a thermal gradient. In fact, any non-homogenous temperature field can be described by multiple thermal gradients, orientated into multiple directions. Thus, for defining the thermal gradient the boundaries of the corresponding measurement area must be defined. The authors defined the ROI by the arteries and veins providing perfusion of the region where the reference is located and by the sensory innervation field of terminal branches of the trigeminal nerve. However, clear boundaries of ROIs are not defined and the thermal gradient, equal to the difference between minimal and maximal temperature within the ROI, remains undefined.

Only the images of 50 subjects without symptoms or history of a cranio-facial disorder were analyzed for normal reference values. Receiver operator characteristics (ROC) analysis was obtained in which temperature dispersion parameter outperformed the best in detecting any reference site. At thresholds of 34.6°C in the frontal view and 34.8°C in lateral views, mean and maximum temperatures performed best with an area under the curve (AUC) of 0.839 for mean and 0.876 for maximum temperature.

In a larger sample, also including 111 subjects with minor symptoms, the side-to-side difference in all 14 pairs of reference sites was in the range between 0.07 and 0.52°C, with minimal differences at the temples and maximal differences at the nasolabial folds. Men presented in each reference site with slightly higher temperatures than women, but at the nasolabial folds, the mean difference was large: 1.1°C.

1.4.5 COMPLEX REGIONAL PAIN SYNDROME

Complex regional pain syndrome (CRPS) is a chronic pain syndrome which affects the upper and lower extremities following injuries, often of minor in nature, but surgery or in rare cases even heart infarction or a lung tumor may be the initial trigger. Early diagnosis and treatment of CRPS improve the functional prognosis. The International Association for the Study of Pain (IASP) classified CRPS into two types as follows: Type I, formerly known as "reflex sympathetic dystrophy," in which a peripheral nerve injury is not observed, and type II, formerly known as "causalgia." Diagnosis of CRPS is based on the "Budapest Criteria," where a temperature difference of 1°C between affected and non-affected limb is required to evidence vasomotor disturbance. There are several non-invasive methods available to detect the components of CRPS. The skin temperature changes can be measured by a thermal camera and a GSR sensor may be used for sweating detection. Quantitative sensory testing identifies sensory deficits, trophic alterations, and edematous swellings found by inspection and palpation. Pain related information is captured in the patient's history and quantified by a pain scale.

Niehof et al. (2006) conducted a study using thermography to evaluate the response to whole-body cooling followed by whole-body warming in patients with unilateral complex regional pain syndrome at the hands in comparison to healthy controls. The authors used a ThermaCAM SC2000 FLIR thermal camera to measure the skin temperature of all fingertips on the dorsal side of both hands. A visual analog scale (VAS) was used to record the pain in the hand region simultaneously with thermal imaging. The authors interpreted the difference in fingertip temperature between the affected and the non-affected hand in patients and between the dominant and the non-dominant hand in controls as a marker of the vasoconstrictor activity. Also, the differences between the minimal fingertip temperature and the maximal fingertip temperature during the whole temperature cycle were calculated in both the involved and contralateral hands. The asymmetry factor, based on histograms of temperature distribution was also used for evaluation. At baseline, the median difference in fingertip temperature was 0.43°C (0.04–0.66°C) in controls and 0.37°C (0.1–0.77°C) in patients. During the total temperature cycle, the average fingertip temperature in controls was a median of 0.95°C (0.50–1.51°C) and in patients, the median was 2.50°C (1.61–3.43°C). Hence, CRPS patients present with a wider response of skin temperature to wholebody thermal stimulation than healthy controls. The best performance as a diagnostic test for CRPS was obtained for the minimum asymmetry factor during the whole temperature cycle with a diagnostic sensitivity of 1 and a specificity of 0.83, followed by the asymmetry factor at rest (sensitivity 1, specificity 0.75). The maximum average fingertip temperature difference during temperature cycle (sensitivity 0.92, specificity 0.75) and the average fingertip temperature difference at rest (sensitivity 0.76, specificity 0.38).

Cho et al. (2016) performed a retrospective study to evaluate the temperature difference distribution between the affected and unaffected limbs in CRPS patients. They observed about 44.3% of patients presenting with 1°C temperature difference between the bilateral limbs. This is not an unexpected finding, because the established diagnostic Budapest criteria, allow the diagnosis of CRPS when the "vasomotor" category is explained by color changes and clinical signs are available in at least two of the four diagnostic categories sensitivity, vasomotor, trophic/motor and sudomotor/swelling. Since temperature was not a predominant sign of CRPS in the investigated population, absence of significant correlations between skin temperature and symptom duration is not surprising.

Radius fracture is a frequent initial injury triggering the development of CRPS. Ammer (1991a), conducted a study on patients, whose plaster fixation was removed due to radiographic signs of bone healing (Ammer, 1991a). Thermal images were recorded immediately after removal of the fixation and a week later, radiography of the distal forearm and the hand skeleton was performed. Radiographs were evaluated for early signs of algodystrophy, which were found in 21 of 41 patients. Mean skin temperature in an area of the wrist and the hand dorsum was based on 10 spot temperatures located at the wrist and the midpoint of each metacarpal bone. Side-to-side temperature difference (affected side minus non-affected side) was calculated. At baseline, the hand temperature on the injury side was 32.8 ± 1.4°C and 31.9 ± 1.7°C on the control side. A week later, the mean temperature of the injured arm decreased to 32.6 ± 1.2°C, while the non-injured control remained at a temperature of 31.9 ± 1.2°C. At baseline, the fractured radius site was in radiographically positive CRPS cases by 1.2°C warmer than the control side. Diagnostic sensitivity, and specificity for radiographic signs of algo-dystrophy was calculated at each time point for a temperature difference threshold of 0.6 and 1.6°C, respectively. At baseline, the diagnostic sensitivity of temperature difference ≥0.6° was 71.4% and specificity was 45%. Increasing the threshold to 1.6°C resulted in a decrease of sensitivity to 33% and an increase of specificity to 95%. Based on these results, it was hypothesized that persistent hyperthermia after bone injuries may promote bone dystrophy in CRPS patients.

Niehof et al. (2008) investigated the discriminative power of the temperature difference between the injured and the non-injured contralateral body part in CRPS patients in comparison to control fracture patients with or without complaints similar to CRPS. All data were acquired after the removal of plaster. There were significantly different temperature findings between CRPS patients, patients with CRPS-like symptoms, and injury patients without complaints. Using the temperature findings of CRPS patients and CRPS-like patients in an ROC analysis at a threshold > 0.99°C, the authors found a sensitivity of 48% and a specificity of 64% for identifying CRPS.

Conwell and Lind (2015) compared the diagnostic accuracy of three infrared imaging methods in a retrospective study with 299 presumptive CRPS patients. Functional infrared imaging is a technique including a cold-water stress of a single lower extremity when CRPS is presumed at the upper extremity and vice versa. The response of the CRPS site and the contralateral uninvolved side is documented by subtracting the image after the cold stress test from a baseline image. While the result in the unaffected limb is a temperature decrease, the response is often a paradox with the increased temperatures at the CRPS sites (Conwell et al. 2010). Functional infrared imaging achieved in the ROC analysis, an AUC of 91.5% (confidence interval 86.9–96.2%), representing a sensitivity of 83.9 and a specificity of 99.2. The AUCs for the quantitative and qualitative analysis were 61.8% (confidence interval 55.2–68.3%) and 53.1% (confidence interval 49.1–57.2%), respectively (Conwell and Lind, 2015).

Overall, the diagnosis of CRPS cannot be based on a single sign such as the temperature difference between the affected and non-affected contralateral body part. Since temperature and other signs of CRPS develop in parallel, temperature observation may be useful for monitoring the course of the disease as suggested in the literature (Ammer, 1991b Schuhfried et al. 2016; Perez-Concha et al. 2020, Huygen et al. 2004, Baic et al. 2017).

1.4.6 FEVER SCREENING

Screening is a different process than finding a diagnosis. Screening is defined as "routine periodical investigation of large non symptomatic strata of the population" (Becker, 2002). Screening procedure results in a high probability of having or not having the disease and must be differentiated from early diagnosis of the disease usually performed after positive screening or after a health check up based on individual request. Thus, a fever screening test aims to separate individuals with a high probability of having an elevated febrile body temperature from afebrile persons. Then, the diagnosis of fever is based on established procedures to detect febrile temperatures. Detected and confirmed fever may itself serve as a screening procedure for infectious, communicable diseases.

However, the results of screening investigations must not be used for the evaluation of the performance of the method applied to screening. Consequently, all cases suspicious and unsuspicious for fever must be checked with correct fever measurements to obtain clear temperature thresholds for the screening procedure. If fever screening is part of a preventive measure against the spread of infectious diseases, the rate of false-negative cases has a higher impact than false-positive cases, because febrile patients due to possible infection are missed if falsely screened as afebrile, thus, increasing the risk of promoting the spread of the infectious agent. Individuals falsely screened as febrile will be detected in the necessary fever measurement as afebrile persons, who bear a minor risk of spreading communicable diseases.

Nguyen et al. (2010) conducted a cross-sectional study by comparing three different infrared thermal detection systems for mass screening of fever detections. The authors used different infrared thermal detecting systems (ITDS) such as FLIR ThermoVision A20M, OptoTherm thermoscreen, and Wahl fever alert imager HSI2000S for capturing thermal images of the face and neck region. They compared self-reported fever and the fever screening results with the oral temperatures obtained with a digital thermometer. Confirmed fever was defined as oral temperatures > 37.8°C, the fever detection thresholds varied between the ITDS: OptoTherm 35.2°C, FLIR 35.8°C, Wahl 21.8°C) The OptoTherm and the FLIR system provided better sensitivity (91% and 90%, respectively) and specificity (86% and 80%, respectively) than the Wahl system for assumed fever cases. The correlation between the screening temperatures and the oral temperature was found to be similar in OptoTherm (r = 0.43) and FLIR (r = 0.42) but significantly lower in Wahl (r = 0.14). The false-negative rates, equal to 1 minus sensitivity, were 9% for OptoTherm, 10% for FLIR, 20% for Wahl, and 25% for self-reported fever.

Zhou et al. performed the fever screening in the facial region using infrared thermal cameras (Zhou et al. 2020) following strictly the recommendations of both the International standard IEC 80601-2-59:2017 and the consensus technical report, ISO/TR 13154: 2017. They acquired infrared thermal images of the face and oral temperatures in a population of 596 subjects. They used two different infrared cameras, a FLIR A325Sc thermal camera and an ICI (Infrared Cameras Inc.) 8640 P-series camera for temperature acquisition in various facial regions and compared them with the sublingual temperature obtained with a digital thermometer and used it as reference T_{ref}. They assessed the temperature in various facial regions such as forehead, canthi, mouth, and entire face. They fixed T_{ref} > 37.5°C as the cut-off temperature for fever detection. They obtained the difference between T_{ref} and temperature of different facial regions as 1.6–2.8°C range for the forehead region, 1.4–2.4°C range for inner canthus, 1.7–1.8°C for the mouth

region. The maximum temperature data at canthi regions produced better performance compared to forehead and mouth regions. Finally, they concluded that
maximum temperature measured in the full-face region could be considered as an
alternative to the recommended inner canthus as the site of temperature measurement in fever screening.

1.5 SUMMARY

Infrared thermography is a non-invasive imaging technique for determining the
surface temperature of an item. Thermal imaging has been utilized to assist in
the diagnosis of a range of medical conditions involving the locomotor and neuromuscular systems. Thermal imaging was initially used to identify breast cancer
in medicine. Thermometry is both a branch of physics dealing with temperature measurement and a science concerned with the manufacture and usage of
thermometers.

Medical thermal cameras are incredibly accurate devices capable of monitoring the temperature of the skin's surface and imaging all areas of the human
body with great precision and clarity. Medical thermography is used in a variety
of settings.

A medical thermal imaging camera generates a thermal image of the human
body that depicts the various degrees of heat that are emitted. A medical thermal
imaging camera can be used to measure the temperature of a specific area of the
body in order to detect abnormally high temperatures caused by inflammation,
circulation problems, injury, or possible infection.

Rapid, non-invasive, and non-contact medical thermography technique.
Medical infrared cameras are a useful supplementary technique for detecting
abnormally warm skin.

When thermal imaging is utilized as a medical device, it can be a very useful
screening tool for determining the temperature of the human body in relation to
potential infection or sickness. This technique is convenient and efficient, and it
may be used in a touchless/non-invasive manner to identify candidates for further
diagnostic testing in a matter of seconds. The use of this screening approach may
help to prevent and slow the transmission of contagious disease, as it allows the
operator to operate the instrument at a wider distance from the patient than forehead-reading thermometers allow. Due to the fact that thermography may be used
to diagnose a wide variety of illness patterns using information about the body's
temperature distribution, it is a highly versatile approach that is very inexpensive
in comparison to other medical engineering processes.

Due to their high mobility, infrared camera systems are particularly attractive
for application. Thermal imaging is used to investigate a wide variety of disorders
in which elevated or lowered skin temperature may indicate the existence of
inflammation in the underlying tissues or increased or decreased blood flow as a
result of a clinical abnormality. Many clinicians now use thermal imaging cameras to diagnose a variety of medical disorders, including arthritis, repetitive strain
injury, muscular pain, and circulatory issues.

REFERENCES

Ammer K Thermographie nach gipsfixierter Radiusfraktur. *Thermologie Österreich* 1991a, 1(1): 4–8.

Ammer K Thermographische Therapieüberwachung bei M.Sudeck. *ThermoMed* 1991b, 7: 112–115.

Ammer K and Ring G. *The Thermal Human Body*. A Practical Guide to Thermal Imaging. Jenny Standord Publishing, Singapore, 2019.

Ammoush M, Gzawi M, Warawreh A, Hijazin R, Jafar H. Clinical evaluation of thermography as a diagnostic tool in oral and maxilla-facial lesions. *Journal of Royal Medical Services* 2018, 25: 45–49.

Anbar M Recent technological developments in thermology and their impact on clinical applications. *Biomedical Thermology* 1990, 10(4): 270–276.

Anbar M Dynamic thermal assessment. In: *Medical Infrared Imaging: Principles and Practices*; edited by Mary Diakides, Joseph D. Bronzino and Donald R. Peterson, CRC Press, 2013, pp. 8.1–8.23.

Baic A, Kasprzyk T, Rżany M, Stanek A, Sieroń K, Suszyński K, Marcol W, and Cholewka A. Can we use thermal imaging to evaluate the effects of carpal tunnel syndrome surgical decompression? *Medicine (Baltimore)* 2017, 96(39): e7982.

Bauer J, Grabarek M, Migasiewicz A, Podbielska H. Non-contact thermal imagigng as potential tool for personalized diagnosis and prevention of cellulite. *Journal of Thermal Analysis and Calorimetry* 2018, 133: 571–578.

Becker N Screening aus epidemiologischer Sicht. *Radiologe* 2002, 42: 592–600.

Biró TS How to measure the temperature. In: *Is There a Temperature? Fundamental Theories of Physics*, (2011) vol 171. Springer, New York, NY. doi:10.1007/978-1-4419-8041-0_2

Chekmenev SY, Rara H, Farag AA. Non-contact wavelet-based measurement of vital signs using thermal imaging In: *The First International Conference on Graphics, Vision, and Image Processing (GVIP)*, Cairo, Egypt, 2005, pp. 107–112.

Cho W, Nahm FS, Choi E, Lee PB, Jang IK, Lee CJ, Kim YC, Lee SC. Multi-center study on the asymmetry of skin temperature in complex regional pain syndrome. An examination of temperature distribution and symptom duration. *Medicine* 2016, 95(52): e5548.

Clay-Warner J and Robinson DT Infrared thermography as a measure of emotion response. *Emotion Review* 2015, 7(2): 157–162.

Cobet R Die Hauttemperatur des Menschen. *Ergebnisse der Physiologie* 1926, 25(1): 439–516.

Conwell TD, Hobbins WB, Giordano J. Sensitivity, specificity and predictive value of infrared cold water autonomic functional stress testing as compared with modified IASP criteria for CRPS. *Thermology International* 2010, 20(2): 60–68.

Conwell TD, Lind KE. Comparison of the diagnostic accuracy of three infrared imaging methods in evaluating patients with presumptive complex regional pain syndrome, TypeI. *Thermology International* 2015, 25(2): 54–63.

Engel J-M, Saier U *Thermographische Standarduntersuchungen in der Rheumatologie und Richtlinien zu deren Befundung*. Luitpold, München, 1984.

Fei J, Pavlidis I, Thermistor at a distance: Unobtrusive measurement of breathing. *IEEE Transactions on Biomedical Engineering* 2010, 988–998.

Garbey M, Sun N, Merla A, Pavlidis I. Contact free measurement of cardiac pulse based on the analysis of thermal imagery. *IEEE Transaction on Biomedical Engineering* 2007, 1418–1426.

Gatt A, Mercieca C, Borg A, Grech A, Camilleri L, Gatt C, Chockalingam N, Formosa C. Thermal characteristics of rheumatoid feet in remission: Baseline data. *Plos One* 2020, 15(12): e0243078.

Giacinto Di A, Brunetti M, Sepede G, Ferretti A, Merla A Thermal signature of fear conditioning in mild posttraumatic stress disorder. *Neuroscience* 2014, 266: 216–223.

Gururajaroa SB, Venkatappa U, Shivaram J M, Sikkandar MY, Amoudi AA. Infrared thermography and soft computing for diabetic foot assessment. *Machine Learning in Biosignal Analysis and Diagnostic Imaging* 2019; 73–97.

Haddad DS, Brioschi ML, Baladi MG, Arita ES. A new evaluation of heat distribution on facial skin surface by infrared thermography. *Dentomaxillo Facial Radiology* 2016, 45: 20150264.

Herman C, Cetingul MP. Quantitative Visualization and detection of skin cancer using dynamic thermal imaging. *The Journal of Visualized Experiments* 2011, 51: e2679.

Houdas Y, Ring EFJ *Human Body Temperature: Its Measurement and Regulation*. Plenum Press, New York, London, 1982.

Huygen FJ, Niehof S, Klein J, Zijlstra FJ Computer-assisted skin video thermography is a highly sensitive quality tool in the diagnosis and monitoring of complex regional pain syndrome type I. *European Journal of Applied Physiology* 2004, 91(5): 516–524.

Ioannou S, Gallese V, Merla A. Thermal infrared imaging in psychophysiology: Potentialities and limits. *Psychophysiology* 2014, 51: 951–963.

Iosif L, Proteasa CT, Magureanu CM, Proteasa E. Clinical study on thermography as modern investigation method for candida-associated denture stomatitis. *Romanian Journal of Morphology and Embryology* 2016, 57: 191–195.

Jakkaw P, Onoye T. Non-contact respiration monitoring and body movements detection for sleep using thermal imaging. *Sensors* 2020, 20: 6307.

Jarlier S, Grandjean D, Delphanque S, Diaye KN, Cayeux I, Velazco MI et al. Thermal analysis of facial muscles contractions. *IEEE Transactions on Affective Computing* 2011, 2(1): 2–9.

Kaspryzk KT, Cholewka A, Balamut K, Kownacki P, Kaszuba N, Kaszuba M, Staneck A, Sieron K, Stransky J, Pasz A, Morawiec T. The applications of infrared thermography in surgical removal of retained teeth effects assessment. *Journal of Thermal Analysis and Calorimetrey* 2021, 144: 139–144.

Kim Y, Park Y, Kim J, Lee EC. Remote heart rate monitoring method using Infrared thermal camera. *International Journal of Engg Research and Technology* 2018, 11(3): 493–500.

Koroteeva EY, Bashkatov AA. Thermal signatures of liquid droplets on a skin induced by emotional sweating QIRT. *Quantitative InfraRed Thermography Journal* 2021, 19(2): 115–125.

Krzywicki AT, Berntson GG, Kane BLO. A non-contact technique for measuring eccrine sweat gland activity using passive thermal Imaging. *International Journal of Psychophysiology* 2014, 94(1): 25–34.

Lewis GF, Gatto RG, Porges SW. A novel method for extracting respiratory rate and relative tidal volume from infrared thermography. *Psychophysiology* 2011, 48(7): 877–887.

Liu C, Van Netten JJV, Van Baal JG, Bus SA, Vander Heijde F. Automatic detection of diabetic foot complications with infrared thermography by asymmetric analysis. *Journal of Biomedical Optics* 2015, 20(2): 026003.

Machin G, Tsai BK. Temperature Fundamentals. In: *Radiometric Temperature Measurements*, edited by Z. M. Zhang, B. K. Tsai and G. Machin, Academic Press, Oxford, UK, 2010, pp. 29–71.

Macianskyte D, Monastyreckiene E, Basevicius A, Adaskervicius R. Comparison of segmented thermal images versus a CT scanning for detection of maxillofacial pathology. *Dentomaxillofacial Radiology* 2019, 48: 20180075.

Nguyen AV, Cohen NJ, Lipman H, Brown CM, Molinari NA, Jackson W, Kirking H et al. Comparison of 3 infrared thermal detection systems and self-report for Mass fever screening. *Emerging Infectious Diseases* 2010, 16: 1710–1717.

Niehof S, Huygen FJPM, Vander Weerd RWP, Westra M, Zijlstra FJ. Thermography imaging during static and controlled thermoregulation in complex regional pain syndrome type I: Diagnostic value and involvement of the central sympathetic system. *Biomedical Engineering Online* 2006, 5: 30.

Niehof SP, Beerthuizen A, Huygen FJ, Zijlstra FJ Using skin surface temperature to differentiate between complex regional pain syndrome type 1 patients after a fracture and control patients with various complaints after a fracture. *Anesthesia & Analgesia* 2008, 106(1): 270–277.

Pauk J, Ihnatouski M, Wasilewska A Detection of inflammation from finger temperature profile in rheumatoid arthritis. *Medical and Biological Engineering and Computing* 2019a, 57(12): 2629–2639.

Pauk J, Wasilewska A, Innatouski M Infrared thermography sensor for disease activity detection in Rheumatoid arthritis patients. *Sensors* 2019b, 19: 3444.

Pavlidis I, Tsiamyrtzis P, Shastri D, Wesley A, Zhou Y, Linder P, Buddharaju P, Joseph R, Mandapati A, Dunkin B, Bass B. Fast by nature-How stress patterns define human experience and performance in dexterous tasks. Scientific Reports 2012, 2: 305.

Pereira CB, Czaplik M, Blazek V, Leonhardt, Teichmann D. Monitoring of cardiorespiratory signals using Thermal imaging: A pilot study on Healthy human subjects. *Sensors (Basel)* 2018, 18(5): 1541.

Perez-Concha T, Tijero B, Acera M, Fernanadez T, Gabilonda I, Gomez-Esteban JC. Usefulness of thermography in the diagnosis and classification of complex regional pain syndrome. *Neurologia* (Engl Ed). 2020 Dec 17:S0213-4853(20)30337-6. English, Spanish. doi: 10.1016/j.nrl.2020.10.011. Epub ahead of print. PMID: 33342641.

Ring EFJ The Thermographic Assessment of Anti-Inflammatory Drug Therapy in Arthritis. M.Sc. Thesis, University of Bath, 1975.

Ring EFJ Technical advances in thermal imaging. In: *Thermal Assessment of Breast Health*, edited by M. Gautherie, E. Albert and L. Keith, MTP Press, Lancaster. 1984, pp. 3–10.

Rossi ABR, Vergnanini AL Cellulite: A review. *Journal of the European Academy of Dermatology and Venereology* 2000, 14(4): 251–262.

Schuhfried O, Herceg M, Reichel-Vacariu G, Paternostro-Sluga T. Infrared thermographic pattern of lower limb complex regional pain syndrome (Type I) and its correlation with pain disease duration and clinical signs. *Physikalische Medizin, Rehabilitationsmedizin, Kurortmedizin* 2016, 26(6): 288–292.

Snekhalatha U, Anburajan M, Sowmiya V, Venkatraman B, Menaga M Automated hand thermal image segmentation and feature extraction in evaluation of rheumatoid arthritis. *Proceedings of the Institution of Mechanical Engineers, Part H: Journal of Engineering in Medicine* 2015, 229(4): 319–331.

Snekhalatha U, Somiya V, Nilkantha G Computer aided diagnosis-based hand thermal image analysis a potential tool for the evaluation of Rheumatoid arthritis. *Journal of Medical and Biological Engineering* 2018, 38: 666–677.

Wesley A, Buddharaju P, Pienta R, Pavlidis I. *A Comparative Analysis of Thermal and Visual Modalities for Automated Facial Expression Recognition*. Advances in Visual computing. Lecture notes in CS. Springer, Berlin Heidelberg 2012 pp. 51–60.

Zhang ZM, Machin G Overview of Radiation Thermometry. In: *Radiometric Temperature Measurements*, edited by Z. M. Zhang, B. K. Tsai and G. Machin, Academic Press, Oxford, UK, 2010, pp. 1–28.

Zhou Y, Ghassemi P, Chen M, McBride D, Casamento JP, Pfefer TJ, Wang Q. Clinical evaluation of fever screening thermography impact of consensus guidelines and facial measurement location. *Journal of Biomedical Optics* 2020, 25(9): 097002.

2 Protocol for Standardized Data Collection in Humans

Following strictly standardized conditions and procedures for capturing infrared thermal images is the mandatory requirement to get valid and reliable information on both image content and temperature estimation. Thermal images of high reliability reduce the workload for image processing to suppress artifacts and correct positioning errors, thus increasing the credibility of automatically obtained temperature readings.

2.1 BACKGROUND AND INTERNATIONAL GUIDELINES

Since all objects with a temperature above zero Kelvin emit some electromagnetic radiation, this radiation can be captured by suitable detectors. The thermal radiation is converted into temperature and displayed as an image representing the site at the object's surface from where the radiation is emitted. In this way, a rough image of the shape of the object is generated, which is determined by the differences in temperature between the object and the ambiance. The texture within a thermal image is also dependent on the contrast between thermal pixels, which represent the anatomical structure of the skin rather poorly. Details of an object, particularly, anatomical information of living beings such as animals or humans are captured due to the fact, that their 3-dimensional form is projected on a 2-dimensional plane (Usamentiaga et al., 2017). A variation in the angle of view of the camera results in a different amount of thermal radiation that can be captured, leading to a temperature contrast that indicates the boundaries of the anatomical regions.

"Directional emissivity" (Wiecek et al., 2000) is one reason why thermography-based temperature measurements are only accurate if taken in a perpendicular view of the thermal imager. The measurement uncertainty of angled camera views is well known since the 1970s (Watmough et al., 1970, Clark, 1976). Recent research reported the influence of angles and distance for thermography-based temperature measurements at the inner canthus of the eye (Vardasca, 2017) and on the Achilles tendon, respectively (Oliver et al., 2020).

Detailed standards for clinical thermography related to equipment, ambient conditions, preparation, and positions of subjects being imaged were developed as late as the beginning of the 21st century, although the American radiologist J.D. Wallace and the English physicist C.M. Cade proposed standardized procedures for clinical thermography in 1975 (Wallace and Cade, 1975). R.E. Woodrough

DOI: 10.1201/9781003245780-2

published in 1982 a book titled *Medical Infra-Red Thermography* focusing on both the clinical technique and applications of infrared thermography, with a detailed discussion of technical aspects of infrared imaging (Woodrough, 1982). Francis Ring at the Royal Hospital for Rheumatic Diseases was the first who studied from 1980 to 1981, the spatial and the temperature resolutions of infrared cameras used in clinical research, opening thereby the field of quality assurance in infrared-based temperature measurements (Ring, 1984). Ring and Ammer summarized the details of the technique of infrared imaging in an article published in 2000 in *Thermology International* (Ring and Ammer, 2000), which became the main reference source for articles discussing factors that influence the information in thermal images (Fernández-Cuevas et al., 2015) or developing a checklist for reporting studies performed with thermography (Moreira et al., 2017). In 2001, a catalog was developed at the University of Glamorgan, proposing body positions for 24 body views for thermal image capture, and 90 regions of interest (ROIs)for temperature measurement (Ammer, 2008a) provides also detailed historical information on developing standard protocols and evaluation of methods used for temperature extraction from thermal images as available at the time of publication. However, the referenced articles cover breast thermography incompletely, although standard positions and views for imaging the female breast were established in the 1970 (Woodrough, 1982, Amalric et al., 1976, 1978).

In 2006, the American Academy of Thermology (AAT) started developing a series of practice guidelines including neuromusculoskeletal (Schwartz et al., 2006), breast (Getson et al., 2015), veterinary (Turner et al., 2019) and dental-oral and systemic health infrared thermography (Schwartz et al., 2015). Unfortunately, these guidelines do not meet the requirements for evidence-based recommendations (Kish, 2001).

In 2017, a checklist was developed in a Delphi process to address conditions that should be reported in thermographic studies since they could affect thermography-based temperature readings (Fernández-Cuevas et al., 2015). Although completeness of reporting of thermographic studies is the primary aim, the checklist confers a great impact on study design since data unavailable due to non-recording cannot be reported. Potential sources of uncertainty are listed in Table 2.1. Each source of variation must be controlled by strictly standardized procedures, which are not yet completely implemented in acquiring and analyzing medical thermal images.

The following sections discuss features of quality assurance in medical infrared thermal imaging.

2.2 THERMAL CAMERA PERFORMANCE CHECK

Within the last two decades, tremendous progress was achieved in the frame rate of acquiring thermal images and their image resolution, but little advances were made in the measurement performance of thermal imagers. Several properties such as the time to stable temperature readings after switching on or the use of

TABLE 2.1

Potential Sources of Uncertainty

- Individual data of participants and their preparation for thermal imaging
- Extrinsic factors such as recent physical activity or physiotherapy
- Wetness of the skin
- Ambient temperature, humidity, and infrared sources in the examination space
- Acclimation time
- Camera type
- Camera settings
- Emissivity
- Size of field of view
- Camera position in relation to the imaged subject
- Image analysis

an external temperature reference already discussed in the 1980s (Woodrough, 1982, Ring, 1984) are still important in the performance assessment of infrared cameras. Camera stability after switching on is an often-neglected information by manufacturers, but a relevant characteristic of infrared imagers employed in regular clinical use. A multi-center study in England monitored the performance of six infrared cameras employed in a study that evaluated patients suffering from Raynaud's phenomenon using infrared thermal imaging (Marjanovic et al., 2018). To examine the stability of each imager during its warm-up period, the blackbody was set to 30°C and left to stabilize for 15 minutes. The thermal camera was then switched on, and images were captured every five minutes for half an hour. All the thermal camera temperature measurements were within 2°C of the actual black-body temperature (30°C), and all camera measurements tended toward 30°C by 30 minutes, apart from the measurements obtained by the camera in one center. Figure 2.1 shows the temperature recorded by each thermal camera in 5-minute intervals from the switch-on.

The drift of radiometric calibration is well known particularly in focal plane arrays (FPA) built with uncooled micro-bolometers since their measurement response depends on both the FPA temperature and the scene temperature (Nugent et al., 2013). There are multiple methods for internal calibration of thermally non-stabilized infrared cameras, but a drift of temperature readings over time remains in the real world of infrared camera applications. Marjanovic et al. reported for various FLIR models, a within-subject standard deviation in the range between 0.06°C and 0.71°C for the baseline and 3-month follow-up thermal camera measurements acquired across all blackbody temperature points (Marjanovic et al., 2018). Poor traceability of calibration or long periods between calibration is associated with large measurement bias to the reference temperatures (Simpson et al., 2006).

Plassmann et al. (2006) proposed some simple tests to check the measuring performance of infrared cameras on a regular basis. Those tests should employ a black body reference source, but the temperature of a water surface provides a

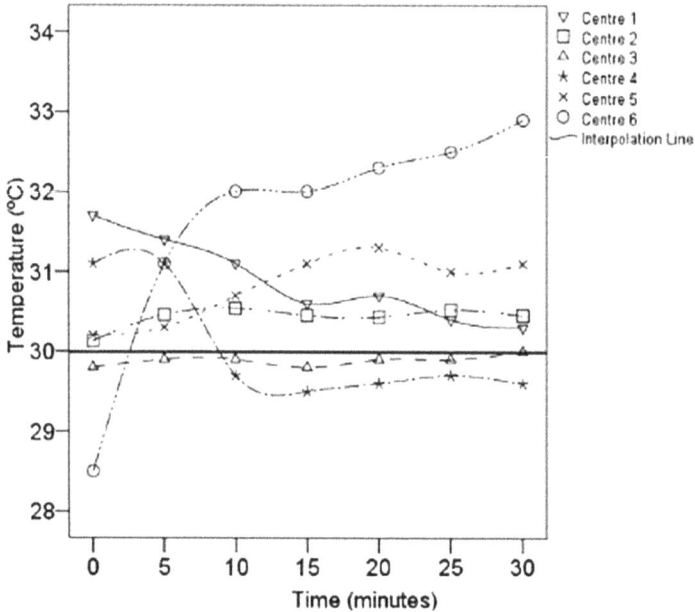

FIGURE 2.1 Camera temperature drift after switch-on. (With permission from: Marjanovic, E.J., Britton, J, Howell K.J. and Murray A.K. Quality assurance for a multi-center thermography study. *Thermology International* 2018; 28(1) 7–13, licensed under a Creative Commons Attribution-NonCommercial-NoDerivatives 4.0 International License. (CC BY-NC-ND 4.0))

good, inexpensive alternative to a black body source. Start-up drift may show a temperature variation of several degree Celsius. For periods of weeks to months, a long-term drift may develop indicated by intra-subject standard deviations between 0.1°C and 0.7°C (Marjanovic et al., 2018). The range of temperature between 15°C and 20°C is the most interesting for medical applications of thermal imaging. The offset variation over this temperature range may be as much as 2°C and different at defined temperature levels (Plassmann et al., 2006). Image non-uniformity can be related to the detector adjustment routines causing variation of temperature readings of single sensors at a uniform reference temperature (Figure 2.2).

Documentation of quality control procedures is important, particularly in case when the infrared camera is part of a regular thermal imaging service. For this condition, the use of a black body reference source is strongly recommended. A logbook that documents the time of the last calibration of the thermal imager at the national standard laboratory should be available on site of the examination room. A schedule of minor standard tests for assessing the performance of the equipment must be part of the logbook. Scheduled tests must be conducted on time and the test results must be recorded accordingly.

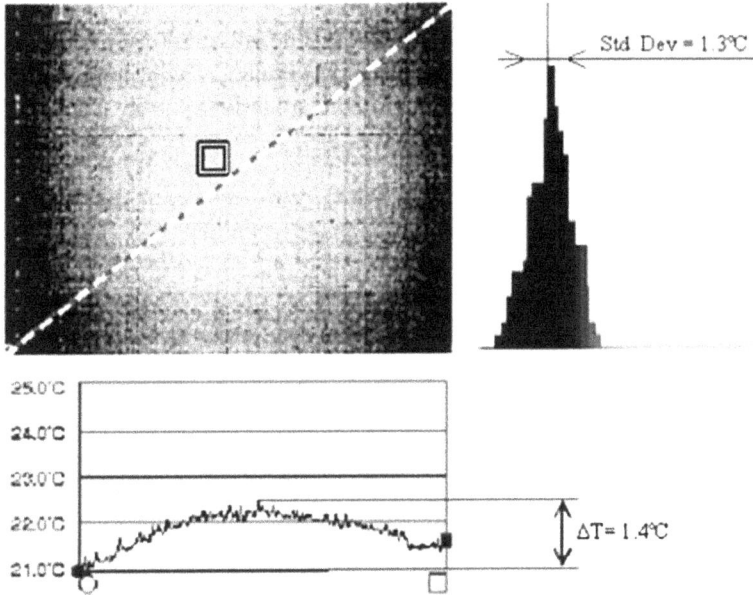

FIGURE 2.2 Non-linearity between opposite corners of a thermal image (dashed line) and pixel noise in a representative area (square). The cross section is shown below the image; the noise level is represented in the histogram to its right. (With permission from Plassmann P., Ring E.F.J. and Jones C.D. Quality assurance of thermal imaging systems in medicine. *Thermology International* 2006, 16, 10–15, with permission from the editor of *Thermology International*.)

2.3 AMBIENT TEMPERATURE CONTROL

According to the second law of thermodynamics, heat flows from high to low levels of temperature, leading to thermal equilibration of objects presenting initially at different temperatures. Thus, the final temperature on the surface of the objects is determined by their starting values. Thermoregulation is a physiological system aiming to maintain a core temperature within a narrow range despite challenges of heat gain from or heat loss to the environment. However, in addition to autonomic thermoregulation, humans extended and modified natural behaviors to achieve thermal comfort, leading already in the early stages of mankind to equivalents of clothing. In the so-called thermoneutral zone, heat exchange with the environment is exclusively controlled via constriction or dilation of cutaneous skin vasculature since neither heat production nor evaporative heat loss is required to keep the inner organs on a constant temperature. Or, in other words, little gains or dissipations of heat may become visible as alterations of local skin temperature. However, not every change in the width of the vascular beds must represent a thermoregulatory response (Marins et al., 2014).

It can be expected that the level of ambient temperature, the size of the skin area exposed to the ambient medium, and the time interval of unclothing affects infrared-based skin temperature readings. Few studies related these theoretical foundations to clinical skin temperature measurements. Khallaf et al. (1994) reported the temperature distribution of the face in seven volunteers dressed in their everyday clothing obtained at two defined room temperatures. Thermal images of the face in the anterior and the lateral view were recorded with an Agema 780M after the subjects had acclimated for 45 minutes to the room temperature of 22°C. The examination room was then cooled to a lower ambient temperature and the subjects were acclimated to this new temperature of 20°C for 40–60 minutes before thermal image capturing was repeated. Before each infrared scan, the sublingual temperature was measured by an electronic thermometer. Spot temperature measurements were made at 12 anatomical sites (Figure 2.3). Interestingly, the decrease in skin temperature varied between the anatomical sites (Table 2.2). Only at measurement point "oral," the mean temperature decay was almost the same as the reduction in ambient temperature. At the site's inner canthus and meatus where maximal temperatures were obtained, the temperature was on average decreased between 0.45 and 0.64°C. The coldest part of the face was the tip of the nose, followed by the ear lobe. The mean drop in nose temperature was 4.1 ± 2°C, while the ear lobe changed only by 1.6 ± 1.3°C.

Different from the interpretation $\bar{t}$ of the authors, these result does not prove that heat removal (=cooling) varies in different parts of the face. Skin temperature represents in a thermoneutral environment mainly the width of cutaneous blood vessels, and relative lower skin temperature is an established sign of vasoconstriction. Thus, Khallaf et al. have nicely shown that the thermoregulatory response to a mild cold stimulus varies in different parts of the face. It is not unexpected that the strongest response became obvious in skin areas rich in arterio-venous-anastomosis (AVA) such as the nose (Bergersen, 1993, Midttun and Sejrsen, 1996)

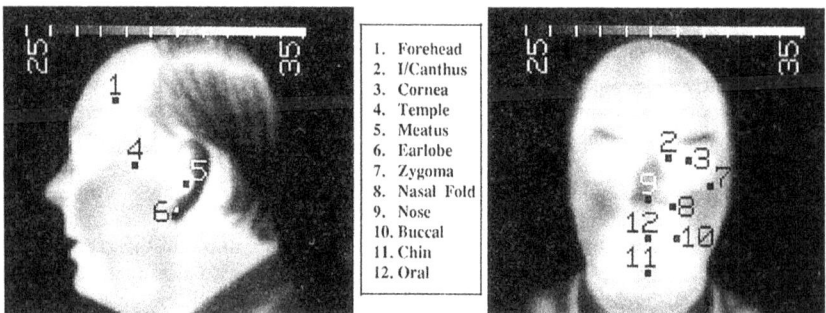

1. Forehead
2. I/Canthus
3. Cornea
4. Temple
5. Meatus
6. Earlobe
7. Zygoma
8. Nasal Fold
9. Nose
10. Buccal
11. Chin
12. Oral

FIGURE 2.3 Thermograms of the face, showing the anatomical points where temperatures were analyzed. (With permission from Khallaf A., Williams R.W., Ring E.F.J. and Elvins D.M. Thermographic study of heat loss from the face. *Thermologie Österreich* 1994, 4 (2) 49–54 with permission from the editor of *Thermologie Österreich*.)

TABLE 2.2

Mean Temperatures of Facial Features at Two Ambient Temperatures

	$\bar{t}$ (°C)	SD	$\bar{t}$ (°C)	SD	$\Delta\bar{t}$ (°C)
Nose	32.58	1.71	28.46	2.01	−4.12
Oral	33.48	1.20	31.51	1.48	−1.97
Nasal Fold	34.07	0.57	32.65	0.57	−1.42
Earlobe	29.34	1.49	28.09	1.64	−1.25
Zygoma	3217	0.71	30.97	1.04	−1.20
Temple	33.55	0.75	32.49	1.00	−1.06
Buccal	33.79	0.55	32.73	0.58	−1.06
Chin	31.81	0.70	30.84	0.86	−0.97
Cornea	33.06	0.53	32.34	0.40	−0.72
I/Canthus	35.08	0.36	34.44	0.22	−0.64
Forehead	33.59	0.45	33.14	0.54	−0.45
Meatus	35.04	0.43	34.64	0.83	−0.40
Sublingual	36.74	0.13	36.49	0.25	−0.25
Ambient $\bar{t}$ (°C)	22.16	0.27	20.27	0.31	−1.89

Source: From Khallaf A., Williams R.W., Ring E.F.J. and Elvins D.M. Thermographic study of heat loss from the face. *Thermologie Österreich* 1994, 4 (2) 49–54 with permission from the editor of *Thermologie Österreich.*

and the ear lobe. These findings point also to the multiple feedback, feedforward, and open-loop components that contribute to thermal balance in the thermoregulatory system operating as a federation of independent thermo-effector loops (Romanovsky, 2007). For example, cold defense loops show individual thresholds of effector activation, allowing a specific response to a range of temperatures The results originating from thermo-effector loops are better explained by the balance points of an interacting network for thermoregulatory responses than by the comparison of temperature input with a setpoint (Imeri, 2017).

Pascoe and Fisher investigated the change in face temperature at sites considered to provide temperature readings that may be used as a substitute for core temperature (Pascoe and Fisher, 2009). Twenty-two college-aged, healthy participants (11 males, 11 females) performed six trials at three ambient temperatures (15.5, 21.1, and 26.6°C) and either 35% or 70% humidity. Participants wore similar clothing in all trials. The trials were performed at the same time each day with participants being equilibrated for at least 15 minutes before temperature measurements were obtained. There was a separation of at least 24 hours between each trial. Core temperature was measured at the rectum, in the esophagus, oral, and at the tympanic membrane. The axilla, the forehead, the temple, and the inner canthi of the eyes were the sites of surface measurements.

Mean rectal temperatures appeared as the most stable and tympanic temperature as the most variable core temperature indicator. Irrespective of ambient humidity, none of the four measurement sites reflected the increment of 6.6°C in room temperature. Axillary temperature was also not much affected by ambient temperature, but forehead, temple, and canthi temperature reflected the altered thermal ambiance nicely as shown in Figure 2.4 and Table 2.3, respectively. The authors stated that their study reinforced the concept that the room environment for thermal imaging must be controlled. They recommended the inner canthi as the most promising site for fever screening with an infrared camera under the condition that the ambient temperature of the examination space is limited to less than 24°C.

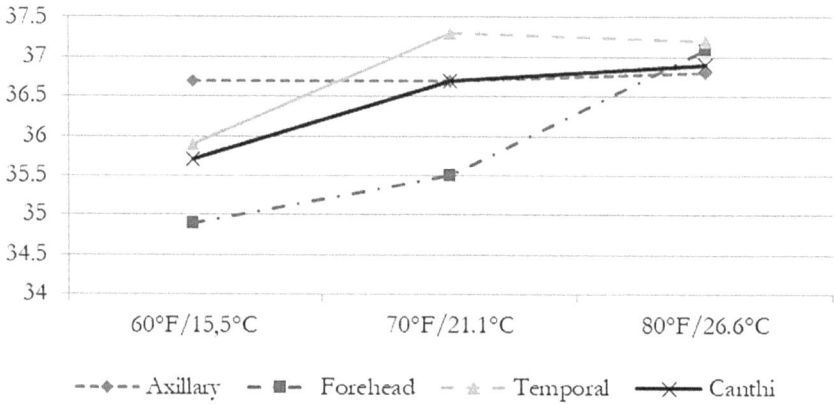

FIGURE 2.4   Axillary and facial temperatures at 70% ambient humidity and room temperatures between 15.5°C and 26.6°C. (With permission from Pascoe D.D. Comparison of measuring sites for the assessment of body temperature: Final amendment. *Thermology International* 2010, 20(1) 36–38 with permission from the editor of *Thermology International*.)

TABLE 2.3

Core and Skin Temperatures at 70% Ambient Humidity and Room Temperatures between 15.5°C and 26.6°C

Measurement Site	Ambient Temperature		
Humidity 70%	60°F = 15.5°C	70°F = 21.1°C	80°F = 26.6°C
Rectal	37.1	37.2	37.0
Oral	36.8	37.0	37.1
Esophageal	36.2	36.3	36.4
Tympanic	36.4	36.7	36.8
Axillary	36.7	36.7	36.8
Forehead	34.9	35.5	37.1
Temporal	35.9	37.3	37.2
Inner canthi	35.7	36.7	36.9

Source: From Pascoe D.D. and Fisher G. Comparison of measuring sites for the assessment of body temperature. *Thermology International* 2009; 19(2) 35–42; with permission from the editor of *Thermology International*.

Removal of clothing disturbs the thermal equilibrium between the skin surface and clothing surfaces, and it requires some time until the skin will become equilibrated with the surrounding air. In Germany, Reeh, Schwan, and Rost followed the idea that the course of temperature change at defined body sites may confer important information about underlying diseases and /or pathologies. Although this "Regulation-Thermography" failed to be accepted by scientific evidence-based medicine, similar periods are described to be necessary for skin temperature stabilization. For the baseline temperature measurement, numerous skin temperature measurements are performed on the head when the patient just entered the examination room, and after undressing to underwear, other body sites are measured. After the first measurement, the patients acclimate to a room temperature of 20–22°C and then the second series of measurements is performed. Rost argues that skin temperature remains stable for 30–40 minutes after the change occurring in the first 10 minutes following undressing (Rost, 1993).

Most guidelines recommend minimal acclimation times of 10–15 minutes (Woodrough, 1982, Ring and Ammer, 2000, Moreira et al., 2017, Schwartz et al., n.d.). Sports scientists from Brazil and Spain observed mean skin temperatures for 20 minutes in a total of 18 ROIs defined in thermal images recorded in 44 college-aged probands of either sex from their upper body and their lower body in an anterior and posterior view (Marins et al., 2014). The reported remarkable temperature difference was observed between the male and female population in the time to reach a statistically significant difference to baseline temperature readings. In men, only the temperatures of the ROIs at the abdomen and lower back presented with a significant change at the end of the observation. Women developed a significant decrease in temperatures at the abdomen, the anterior thighs, the lower back, the dorsal hand, and the dorsal forearm. At the abdomen and the lower back, a significant temperature fall was detected after 14 minutes of observation, and thigh temperature decrease was observed 16 minutes after baseline readings. At hands and forearms, a significant difference to baseline readings was found at the end of the observation time. The final dorsal hand temperature was also significantly different from the temperatures obtained at 2 and 6 minutes, while the final temperature of the dorsal forearm was significantly lower than the temperatures recorded at 2, 4, 6, and 8 minutes. The authors interpreted their findings that for analyzing whole body thermograms, an acclimation time of 10 minutes is sufficient for both sexes, although some body parts may equilibrate in a shorter period, particularly in men.

When entering from extreme outside temperatures, a 30-minute adaptation in a waiting room was recommended prior to the period of acclimation to the temperature of the examination space (Engel et al., 1985). This recommendation is supported by a study conducted in Auburn, 20 years after this recommendation was published (Fisher et al., 2008). David Pascoe's group acclimated 6 male and 11 female college-age participants for 60 minutes to a room temperature of 24.2°C wearing minimal attire to expose maximal skin surface to room temperature. Then, the participants were exposed for 20 minutes to either a hot (31.7°C) or a cold (18.9°C) environment and returned to the room temperature of 24.2°C for another 60 minutes. The thermal images of the anterior superior, anterior inferior, posterior superior, and posterior inferior view were taken at the end of the initial 60 minutes equilibration period, and immediate-post, after 15-, 30-, 45-, and 60-minutes periods after cold or hot exposures. Mean peripheral skin temperature decreased from baseline readings of 29.7–28.5°C following cold exposure. Peripheral skin temperature remained significantly different from baseline values until the 30-minute recovery time. Following the hot trial, mean peripheral skin temperature increased from baseline readings of 29.6–31.0°C, remaining significantly different for the entire 60-minute observation time. The authors suggested that a 30-minute equilibration period may be more reliable for ensuring proper equilibration, as opposed to the previous standard of 15 minutes.

Fever screening using an infrared radiometer may also require prolonged acclimation time to room temperature if outside temperatures are below 0°C. Under the condition of outdoor temperatures between −5°C and 0°C, core temperature is regularly underestimated when based on forehead temperatures recorded with an infrared thermometer immediately, 1, 3, or 5 minutes after entering the hospital (Dzien et al., 2021).

2.3.1   CHOICE OF AMBIENT TEMPERATURE

A thermal image of any object can only be seen when the surface temperature at the object's edges differs from the background temperature. Thus, an ambient temperature between 18 and 26°C should produce a sufficient thermal contrast to the human skin.

A room temperature of 20 ± 0.5°C may be appropriate for expected local hyperthermia because the heat preserving reaction of partly uncovered skin to this slightly cool environment will increase the contrast between unaffected skin with slightly constricted vessels and the dilated vasculature of the affected skin.

A room temperature of 24 ± 0.5°C may be appropriate for expected local hypothermia since no heat dissipating or preserving reaction of partly uncovered skin will occur in this almost indifferent thermal environment.

However, in the examination of totally undressed, naked subjects, the room temperature should be 22 ± 0.5°C for suspected inflammatory conditions and 26 ± 0.5°C for the identification of reactive temperature changes.

The threshold for evaporative skin cooling due to sweat secretion may vary between people acclimatized to either a warm or a cold climate. Thus, an individual acclimatized to cold climate may start sweating at ambient temperatures whereas a person acclimatized to a hot tropical climate may not respond with thermoregulatory sweating.

2.3.2   STRUCTURE OF THE EXAMINATION SPACE

Thermal images obtained under uncontrolled conditions cannot be analyzed in a scientific way, since they have anecdotical value at best. A major source of uncertainty in both the image content and thermometry results is the location where thermal images are recorded. Thus, infrared thermal images used in medicine must be recorded under controlled conditions in a specially designed examination space. The room should have a minimum size of 2 × 3 m, but larger rooms are preferable. The budget for estimating the least distance in one direction includes the following:

- 30 cm distance from the wall to the subject to be imaged plus
- 30 cm distance between the wall and camera operator plus
- 1 m distance behind the camera to operate the camera and its controlling computer plus
- The distance between the camera and an individual to take an image of the upper or lower part of the human body or an object of 1.2 m height. For total body thermograms, the distance must be calculated for an object of 2.2 m height. This distance portion is determined by the optical features of the lens such as the field of view.

Other important structural features of the examination room are:

- Blacked out or obscured windows with double glazing.
- Fluorescent not tungsten or halogen lighting.
- Non-reflective surfaces where possible. Since emissivity and reflectivity are surface properties, surface roughness is of higher importance than surface color. A shiny, and smooth, black-colored surface reflects more incident radiation than a mat, slightly rough surface in white color.
- Water supply and a wash basin are helpful when cold water provocation should be conducted.

Temperature control of the examination room is a prime requirement. The air conditioning equipment should be located so that direct draughts are not directed at the subject being imaged, and that overall air speed is kept as low as possible (less than 0.3 m/s). Air flow directed to the hoofs of a horse at speeds ranging from 0.5 to 1 m/s over 1.3–2 to 3–4 m/s affected the temperature reading from thermograms recorded in this condition (Westermann et al., 2013). A suspended perforated ceiling with ducts diffusing the air distribution evenly over the room at air-flow speeds between 0.1 and 0.13 m/s is ideal. Calculation of heat load per room/person was proposed (Love, 1985). The air conditioning system should stabilize room temperature for 1 hour within ± 0.5°C.

Room temperature and relative humidity must be displayed in the examination space. Large display digital devices are ideal. Their sensors should be located near the subjects being imaged, but at least 0.3 m away, and clear of walls and surfaces and draughts of air conditioning.

A patient cubicle, large enough for undressing/waiting during acclimation must be located within a temperature-controlled area. Suitable cloth/paper mats avoid contact with the floor. A place for jewelry, watches, etc., should be provided.

A height adjusting stool without back support should be available for imaging the individual in a sitting position. This stool may also be used for the position required for the view "lateral leg." If the ankle joints must be imaged in a perpendicular view, a low stage or two-level staircase is required to move the subject to the correct level.

2.4 SUBJECT SELECTION

The purpose of thermal imaging must clearly be defined. In clinical trials, the study aim may be the evaluation of the diagnostic utility infrared thermography or the detection of the effects of any interventions based on changes obtained in thermal images. In the case of an imaging service for physicians, the clinical question of the referring doctor defines the conditions for recording and analyzing the thermal images.

The basic requirement for investigating the diagnostic utility is experimental data on the performance of an infrared camera as both a measurement and an imaging device. The uncertainty of temperature measurements caused by the camera's hardware and software must primarily be determined against a well-controlled external temperature source. Image characteristics must be compared to other imaging techniques providing an image of the object's surface.

Since the surface temperatures of humans depend grossly on the controlled core temperature and the ambient temperature, they cannot be considered as random variables. Thus, absolute temperatures are of minor diagnostic interest, but relative temperatures such as the difference in temperature of defined points or areas are valid descriptive parameters of temperature distribution. The determination of "thermal symmetry," a high similarity of mean temperatures in measurement areas located in corresponding anatomical regions, became the first quantitative method to analyze clinical thermal images. However, the generation of reference values of relative temperatures of the human skin requires several controlled conditions, which were not always met in studies conducted in the past. The deficits include absent definitions of being completely healthy or having a disease, incomplete reporting of study population recruitment, of body positions and camera views for image capturing, and of details of image analysis. It is therefore not unexpected, that relative temperatures vary between anatomical regions in between $0.0 \pm 0.1°C$ (Zhang et al., 2019) and $1.0 \pm 0.1°C$ (Goodman et al., 1988). A study from South Korea reported for 934 normal subjects measured absolute and relative temperatures from thermal images of the trunk, the upper, and the lower extremities. Measurement uncertainties were calculated in relation to the ROI, room temperature, and camera performance separately for sex and membership in four age classes. The results were illustrated by the example of the temperature of the dorsal hand in men aged 40–49 years. While the combined standard uncertainty of absolute hand temperature was 6.4°C, uncertainty was only 0.7°C for the temperature difference between the right and the left hand, and expanded uncertainty was 1.45°C.

If studies on the diagnostic utility of thermal imaging are based on temperature measurements, the threshold of asymmetry should be between 0.7 and 1.5°C even when the statistical evaluation of a study sample favors a smaller value. The diagnostic accuracy must be derived from the comparison of temperature asymmetry with the established gold standard for diagnosing the disease under investigation. Correlations between temperature values and other physiological parameters do not confer information on correctly or falsely identified cases.

Studies on the employment of thermal imaging as an outcome measure require the definition of the accuracy of thermal images to identify clinical endpoints. A clinical trial endpoint was defined as a characteristic or variable that reflects how a patient feels, functions, or survives. Thus, clinical endpoints are distinct measurements or analyses of disease characteristics observed in a study or a clinical trial that reflect the effect of a therapeutic intervention (Biomarkers Definitions Working Group, 2001). Surrogate endpoint is another important term defined as a marker that is intended to substitute for a clinical endpoint. Exploration of thermal imaging as an outcome measure is the evaluation of whether thermal images can serve as a surrogate endpoint, which is expected to predict clinical benefit (or harm or lack of benefit or harm) based on epidemiologic therapeutic, pathophysiologic, or other scientific evidence. The use of thermal images as surrogate endpoints in a clinical trial requires the specification of the clinical endpoints that are being substituted, the class of therapeutic intervention being applied, characteristics of the population, and the disease state in which the substitution is being made (Biomarkers Definitions Working Group, 2001).

A mandatory property of an outcome measure is its sensitivity to change or, in other words, its ability to detect and quantify a true, non-random change. Determination of the sensitivity to change is often difficult due to the absence of a criterion standard for change in health status (Stratford and Riddle, 2005). Based on the composition of the investigated population with respect to their change behavior, different statistical approaches exist to quantify the amount of change. However, the obtained coefficients may be contradictive when all have been applied to the same study population. Observing the change in the intervention under investigation during a pilot study was recommended to overcome the problem of an undefined standard criterion for change.

Sufficient preparation of the persons being imaged is extremely important. Preparation includes the following:

1. information on the procedures performed during the examination,
2. instructions on the behavior prior to the thermal imaging and
3. collection of biographic and other personal data at the beginning of the imaging session.

2.4.1 PROCEDURES PERFORMED DURING THE EXAMINATION

Patients must be informed on the potential and purpose of thermal imaging. They must know that they have to undress to a certain degree and wait in this clothing condition in a room with temperature controlled to 24°C for 15 minutes, in a defined posture/body position and with as little movements as possible. Body parts intended for imaging should not have contact with furniture or other body parts. For example, if the buttocks should be imaged, the patients should stand and not sit. Prior to the capture of facial thermograms, touching the face with the hands should be avoided. Following the waiting period, thermal images of their total body/defined anatomical regions will be captured in defined body positions and camera views.

2.4.2 BEHAVIOR PRIOR TO THE THERMAL IMAGING

Several activities and conditions performed up to 24 hours prior to thermal imaging may affect the temperature distribution. Some, but not all the following recommendations are supported by experimental results. Instructions include avoiding alcoholic beverages, smoking, caffeine, large meals, ointments, cosmetics, and showering for four hours before the assessment. Also, sunbathing (e.g., UV sessions or direct sun without protection) should be avoided before the assessment (Moreira et al., 2017). Unless extrinsic factors affecting skin temperature represent the research question under investigation, interventions such as physical exercise, massage, electrotherapy, ultrasound, heat exposure, or cryotherapy should be avoided 24 hours prior to thermal imaging.

2.4.3 DATA CHECKING AT THE BEGINNING OF THE IMAGING SESSION

Check personal data of each person being imaged. If the individual participates in a clinical trial, check if the schedule of the examination is timely met and what thermal images the study protocol is requiring. Thermal imaging studies in the field of sports and exercise science should include information on body mass index (BMI) and a measure of physical activity frequency (hours/day of training) and characterization (activity description).

In case of referral from a physician, check if a clinical question is formulated, and indicate the anatomical region of which thermal images should be recorded (Biomarkers Definitions Working Group, 2001). The minimum medical information combined with the request for thermal imaging should include the following:

- The patient's name, gender, date of birth.
- Their affected body region and symptoms such as pain, paresthesia, swelling, redness, heat sensation, muscle weakness.
- Diagnosis (suspected or already established).
- Recent or previous injuries in the symptomatic body region.
- Recent surgery.
- Current treatment with drugs and/or physiotherapy.

Get verbal confirmation that instructions on avoiding distinct behaviors have been followed. Any condition that could not be avoided should be reported. The use of any medicinal treatments or drugs should be recorded.

2.5   PATIENT POSITION AND IMAGE ACQUISITION

For getting accurate and reliable thermal images, the camera position in relation to the object is the critical issue. Since the human body has a curved surface, any deviation from a perpendicular view results in a parallax error in the captured image. Thus, fixation of the thermal imager allowing only horizontal or vertical view is helpful in avoiding parallax errors. The camera should be mounted to a photographic stand that can quickly adjust the level to the height of the body ROI. Since the lowest possible level is at a 20–30 cm distance to the floor, it is recommended to place subjects on a low stage or two-level staircase when the rear foot or the anterior surface of the ankle joint should be imaged. Lower body parts can, of course, be imaged in an angled view, but their temperature readings must be corrected if compared to temperatures obtained from thermal images captured in a perpendicular view.

Figure 2.5 illustrates the effect of perpendicular and angled views on the back of the foot. The mean temperature at the back of the forefoot is 0.9°C higher in the perpendicular view than in a view angle between 24° and 45°.

Different from radiography, an agreed catalog of standard views for imaging defined anatomical regions does not exist. Ammer developed in 2001, a catalog of 24 camera views and related body positions as the framework for an atlas of thermal images of healthy subjects. The image content was defined by aligning anatomical landmarks to the edges of the image achieving thereby a normalization of anatomical regions by their body proportions.

Figure 2.6 illustrates the principle of producing this catalog at the view "face." The distance between the most cranial point of the head and below the chin at the level of the hyoid bone is the measure for normalization. Other conditions such as a vertical position of the head, neither rotated nor tilted to the side define the

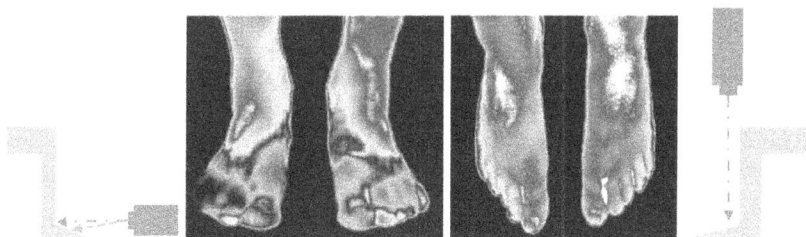

	Right forefoot	Left forefoot
Angle of view 24°-45°	28.8 ± 0.4 °C	28.3 ± 0.3 °C
Perpendicular view	29.7 ± 0.01°C	29.2 ± 0.2 °C

FIGURE 2.5   Temperature at the back of the forefoot in dependence of the angle of view.

upper edge of the image:
the most cranial point of the head
lower edge of the image
below the chin at the level of hyoid bone
other conditions
head is a vertical position, neither rotated nor tilted to the side

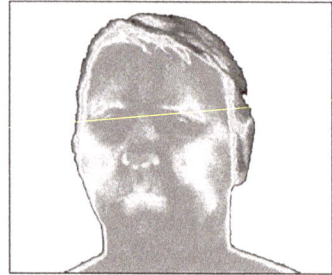

FIGURE 2.6   View 4: Face (anterior view); Code: FA. (With permission from: Ammer K. The Glamorgan Protocol for recording and evaluation of thermal images of the human body. *Thermology International* 2008, 18: 125–144, with permission from the editor of *Thermology International*.)

upper edge of the image:
iliac crest
lower edge of the image:
sole
other conditions:
hip and knee are approximately 90° bent, foot is placed on a chair, the total leg is in the sagittal plane of the body

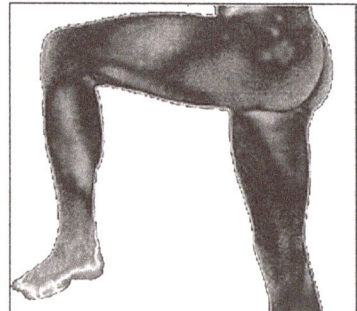

FIGURE 2.7   View 18: Leg, left (lateral view); Code: LLL. (With permission from: Ammer K. The Glamorgan Protocol for recording and evaluation of thermal images of the human body. *Thermology International* 2008, 18: 125–144, with permission from the editor of *Thermology International*.)

standard position of the head. This approach guarantees that irrespective of the body height, each depicted body part shows little variation in the targeted body dimension.

Another principle was that the view should allow imaging as much as possible of the surface of an anatomical region in a perpendicular view in a body position that can easily be reproduced because all angles between large joints of extremities are either 90° or 180°. Figure 2.7 shows this principle in the lateral view of the leg. The leg is flexed in hip and knee, 90°, and the angle between the shank and the sole of the foot is also 90°. To achieve this, the foot is placed on a height adjusting supporting surface to stabilize this position of the total leg in the sagittal plane of the body. The measure of normalization is the distance between the tangents through the iliac crest on the upper edge of the image and through the planta of the foot on the lower image edge.

The bias from the recommended position was determined in 10 views by the number of pixels between the landmark and the image edges. The largest distance to the upper image edge was found in the view "both ankles anterior" with a 95% confidence interval between 134 and 184 pixels. The view "face" was closest to the edges with a distance to the upper edge in the estimated range of four pixels across 95% of cases and eight pixels in the 95% confidence interval for the distance to the lower image edge (Ammer and Ring, 2019, Ammer and Ring, 2008).

In an instructional course on medical thermography, the hands of two volunteers were imaged. After allocation into two groups of seven students, each group member was instructed to follow the instructions for standard views used in the protocol of the University of Glamorgan. Each course participant then recorded two separate thermal images of the dorsal hands of one volunteer in a perpendicular view. The repeatability of the position of hands was evaluated by the difference in the area covered by the hands. In thermal images with 89600 pixels, the mean hand size varied in repeated image capture in one group by approximately 2300 pixels, and by 600 in the other group. However, there were individual deviations from the mean hand size in the range of 5000–7600 pixels (Ammer and Ring, 2004).

Averaging the thermal images recorded in a specific view also indicate the deviation from the position defined in the Glamorgan protocol. Using overlaid contour outline masks can reduce the error in positioning (Plassmann and Jones, 2003). Such masks can be generated from averaged views but must be adjusted with respect to the given definition. Including predefined ROIs in these masks also improves the reproducibility of temperature readings.

2.5.1 STATIC OR DYNAMIC THERMAL IMAGING

Capturing once a single thermal image in a defined view is a typical example of static thermography, documenting the temperature distribution on the surface at the moment of image capture. Static thermal images recorded hours to days later, document the temperature distribution at the later time point and may show a change in the temperature map.

Taking images at a shorter interval allows to detect temperature changes over time, transforming static to dynamic thermal imaging. Intervals between single images range from 1 s to 5 minutes. Thermal videos may record images at frame rates up to 30 Hz. Dynamic thermal imaging at frame rates between 0.5 and 30 Hz can detect cyclic changes in temperature associated with other physiological functions such as heart rate or breathing rate.

The term dynamic thermal imaging is often restricted to image series recorded after the application of thermal stimuli such as convective fan cooling or immersion of body parts into a cold-water bath. Fan cooling originally used to enhance thermal contrast in breast thermography, attracted new interest as a method to detect in thermal images perforating vessels of the skin, which may serve as a nourishing vessel for skin flaps (Kolacz et al., 2017, Weum et al., 2016). Detection of perforator vessels is based on the qualitative evaluation of thermal images; however, fan cooling is performed in a standardized manner, and first appearing hot spots were always associated with arterial Doppler sounds (Weum et al., 2016).

There is a long history of the cold-provocation test to detect a suspected Raynaud's phenomenon by thermal imaging. However, the idea to detect finger color changes following cold exposure goes back to the British cardiologist Sir Thomas Lewis. Francis Ring proposed in 1980, a combined thermal gradient by summing up the distal–dorsal temperature difference prior and 10 (15 or 20) minutes after the cold challenge for assisting the diagnosis of Raynaud's phenomenon (Ring, 1980). He also reported the consensus of a panel of thermal imaging experts on the thermographic assessment of patients with Raynaud's phenomenon in 1988 (Ring et al., 1988). Ammer tried to extract from the thermographic literature, procedures which address patient's preparation, temperature of the examination room, temperature and duration of the immersion bath, position of hands, time of follow-up after the cold challenge, and method of evaluation. The review article reported a wide variation in water temperature of the immersion bath, also of the duration of immersion. More than 20 different methods for evaluation of hand temperatures were published (Ammer, 2009). Applying a mild cold stress with the hands covered with plastic gloves being immersed in water of 20°C for 1 minute is in the author's experience, an easily performed provocation test for Raynaud's phenomenon. After emersion, thermal images recorded for 20 minutes at 5-minute intervals allow to determine combined thermal gradients. This kind of analysis can be performed semi-automatically with dedicated software (Plassmann and Jones, 2003).

The diagnostic value of the cold challenge was questioned (Pauling et al., 2011, Herrick, and Murray, 2018). A multi-center study investigated the validity and reliability of responses to hand cold challenge as measured by laser speckle contrast imaging (LSCI) and thermography (Wilkinson et al., 2018). The test–retest reliability was determined for the distal dorsal difference (DDD) at baseline, i.e., prior to the cold challenge, the area under the curve (AUC) of the reperfusion/rewarming curve, and the maximum blood flow rate/skin temperature after rewarming (MAX). Based on intraclass correlation coefficients (ICC), test–retest reliability LSCI was substantial for DDD, AUC, and MAX, while thermography reliability of AUC and MAX was also substantial, and almost substantial for DDD. Convergent validity between LSCI and thermography was derived from estimates of the latent correlation coefficient rho, which would be equal to 1, if the techniques measured the same construct. Rho values were 0.65 for DDD, 0.87 for MAX, and 0.94 for AUC, indicating similar performance of LSCI and thermography in detecting post-cold-challenge rewarming or reperfusion, respectively.

A recent study from Italy that applied a deep learning algorithm also reported good performance of baseline temperatures in discrimination of healthy subjects and patients with primary or secondary Raynaud's phenomenon (Filippini et al.,

2021). However, using the DDD only at baseline and not at the end of observation after cold challenge may miss up to 20% of Raynaud's phenomenon patients (Ammer, 2013).

2.6 THERMAL IMAGE ANALYSIS

In each view of the Glamorgan protocol, one to four ROIs are defined. Figure 2.8 shows the ROIs of the view "face." Shape and alignment to anatomical landmarks are defined for each ROI.

Positioning of ROIs and their temperature readouts can easily be reproduced with a small mean uncertainty of a group of evaluators, but with large errors in repeated analysis by individual observers (Ammer, 2006). The ROI "half of the face" can be used to check the rotation deviation of the head. In case of correct rotation position, the flipped ROI of one side perfectly matches the other half of the face. Normally, the face presents with a high grade of thermal symmetry and a side-to-side difference in temperature greater than 0.15 between the two halves of the face can be regarded as abnormal (Ammer et al., 1997). This does not mean that regional temperature differences of the face do not exceed this low threshold, but the temperature distribution in adjacent facial regions may correct the disturbed balance, thus resulting in high similarity of mean temperatures of facial halves.

Number of ROIs: 4
ROI 1: right half of the face (**red**)
Shape: polygon
Outline of the face from the scalp to the chin, following the midline of the face
ROI 2: left half of the face (blue)
Shape: polygon
Outline of the face from the scalp to the chin, following the midline of the face
ROI 3: right half of the forehead (**yellow**)
Shape: rectangle
Lower edge: adjacent to the right eyebrow
Right upper corner: adjacent to the hair line
Left upper corner: adjacent to the midline of the face
ROI 4: left half of the forehead (green)
Shape: rectangle
Lower edge: adjacent to the left eyebrow
Left upper corner: adjacent to the hair line
Right upper corner: adjacent to the midline of the face

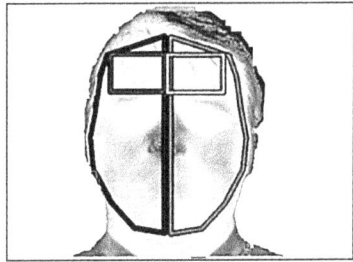

FIGURE 2.8 View "face, code: FA. (With permission from: Ammer K. The Glamorgan Protocol for recording and evaluation of thermal images of the human body. *Thermology International* 2008, 18: 125–144, with permission from the editor of *Thermology International*.)

Skin temperature distribution is characterized by thermal symmetry between both sides of the body and thermal gradients from central to periphery manifesting in higher temperatures of cranial or proximal anatomical regions compared to caudal and distal body parts. Disturbances of both the side-symmetry and the proximal–distal temperature gradients are the basis of quantitative interpretation of infrared thermal images in medicine. A higher foot than knee temperature in diabetics (Melnizky et al., 2002) or lower fingertip than dorsal hand temperature are examples of altered proximal–distal thermal gradients. However, typical thermal gradients may be completely removed in case of long-standing severe motor deficits. Thresholds for abnormal temperature differences have been discussed above in Section 2.4.

A common recommendation for quantitative analysis of thermal images is to extract minimum, maximum, and mean temperature ± standard deviation from the ROIs. Despite identical extreme and average temperatures of three ROIs, their dispersion of temperature values may be completely different ranging from normal over skewed to bimodal distributions (Ammer and Formenti, 2016). Histogram-based analysis can be used to discriminate data with similar dispersion measures, but their relationship to clinical endpoints is not yet established. Plassmann recently showed that based on histogram analysis of liquid crystal thermograms of the foot, the discrimination of stable functional health conditions of the foot from deterioration may be possible (Plassmann and Kluwe, 2021).

Manual definition of ROIs is the gold standard to which automatic definition of measurement areas must be compared. Although the upper range of temperatures within a thermal image can easily be detected by modern algorithms, they are of diagnostic value only when located within the anatomical ROI. Due to their smaller uncertainty (Zhang et al., 2019), obtaining relative temperatures is more relevant than estimating absolute temperatures, because a thermal imager is a thermometer of quite a large measurement uncertainty. Image processing may add further uncertainties when properties of centigrade as an interval scale are neglected. Possible solutions to this problem include combining image segmentation for the definition of ROI with the original unprocessed temperature map. Defining bands of low, moderate, and high temperatures as proposed for the evaluation of breast thermograms is also an alternative solution (Madhu et al., 2016).

Hot spots are easily detected in the upper range of temperatures. A common definition is a small area whose mean temperature is different by a threshold, typically 0.5°C or 1°C, from the surrounding temperature. Such hot spots at the elbow predict pain elicited by firm pressure or resisted movements and also thresholds of algometry (Ammer, 1995). Hot spots may also coincide with tender points of fibromyalgia patients (Ammer, 2008b). In these patients, hot spots may not always be identified in the upper range of temperatures, although the condition of presenting with a higher temperature than the surroundings is met. A previous study applied the features of the CTHERM software, to identify hot spots in thermal images obtained from the back of fibromyalgia patients. Continuous and stepwise color scales were used in thermal images with compressed and uncompressed color scales. Sets of two or three isotherms were defined as an alternative to the visual identification of hot spots when combined with compressed color scales. The procedure conducted with three isotherms was as follows: The first isotherm with the highest threshold was set when at least one spot was enclosed in the upper range of temperature, Then, three isotherms, $0.5°$ apart were generated. Moving such a set over the total range of temperature within a thermal image can easily detect hot spots as the temperature difference between the 1st isotherm with the lowest threshold and the third isotherm with the highest threshold is at least equivalent to the chosen temperature between the isotherms. Figure 2.9 shows an explanatory result of this procedure.

The number of hot spots varied between the different methods of identification, with best results in color-compressed images and isotherms (Ammer, 2011). Reliability for repeated evaluation based on ICC was substantial for isotherms and uncompressed images and excellent for compressed images. An ICC value of 86% was detected between the first count in compressed images and the first count in isotherms. An interesting observation is that some hot spots are identified at the same sites assumed to be tender in fibromyalgia patients, while other hot spots are suggestive of myofascial trigger points. Software that can automatically detect hotspots at various thresholds to the thermal surrounding would help to understand the role of thermal images in the assessment of myofascial pain syndromes. Combining identified hot spots with maps of tender points or trigger points in superficial muscle layers would be a big step toward a computer-assisted diagnosis system of myofascial pain based on thermal imaging in combination with other assessment tools.

In conclusion, standardization of thermal imaging must be applied beyond quality assurance procedures of the thermal imager. Features of the imaged human subject are a major source of both measurement uncertainty and variation in image quality. Defining of measurement area must be meaningful for medical interpretation, and image processing and automatic analysis must not alter the radiometric information contained in thermal images.

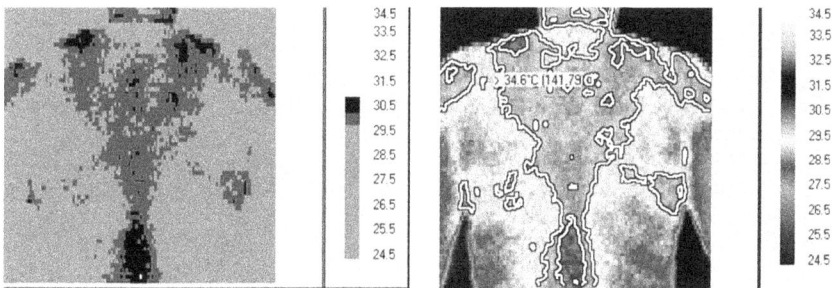

FIGURE 2.9 Thermal image color-coded hot spots and the same image with a set of 3 isotherms. (With permission from: Ammer K. Repeatability of Identification of Hot Spots in Thermal Images is influenced by image processing. *Thermology International* 2011, 21(2) 40–46, with permission from the editor of *Thermology International*.)

REFERENCES

Amalric R, Giraud D, Altschuler C, Deschanel J, Spitalier JM Analytical, synthetic, and dynamic classification of mammary thermograms. *Acta Thermographica* 1978, 3: 5–17.

Amalric R, Giraud D, Altschuler C, Spitalier JM Value and interest of dynamic telethermography in detection of breast cancer. *Acta Thermographica* 1976, 1: 89–96.

Ammer K Thermal evaluation of tennis elbow. In: Ammer K., EFJ Ring (eds), *The Thermal Image in Medicine and Biology*. Wien: Uhlen Verlag, 1995, pp. 214–219,

Ammer K Repeatability of temperature measurements at the forehead in thermal images from the standard view "face" *Thermology International* 2006, 16(4): 138–142.

Ammer K The Glamorgan Protocol for recording and evaluation of thermal images of the human body. *Thermology International* 2008a, 18: 125–144.

Ammer K Thermal imaging: A diagnostic aid for fibromyalgia? *Thermology International* 2008b, 18(1): 45–50.

Ammer K Cold challenge to provoke a vasospastic reaction in fingers determined by temperature measurements: A systematic review. *Thermology International* 2009, 19: 109–118.

Ammer K Repeatability of identification of hot spots in thermal images is influenced by image processing. *Thermology International* 2011, 21(2): 40–46.

Ammer K Temperature of the fingertips in subjects with suspected Raynaud's phenomenon. In: Nowakowski A, Mercer J. (eds) *Lecture Notes of the ICB Seminar*. Springer, Warsaw, 2013, 94, pp. 95–98.

Ammer K, Formenti D Does the type of skin temperature distribution matter? *Thermology International* 2016, 26(2): 51–54.

Ammer K, Melnizky P, Schartelmüller T Thermographie bei Fazialisparese *ThermoMed* 1997, 13: 6–11.

Ammer K, Ring EF Repeatability of the standard view of both dorsal hands. Results from a training course on medical infrared imaging. *Thermology International* 2004, 14(3): 100–103.

Ammer K, Ring EFJ. Standard procedures for infrared imaging in medicine. In: Diakides NA and Bronzino JB (eds) *Medical Infrared Imaging*. Boca Raton: CRC Press 2008, 22, pp. 1–122.14.

Ammer K, Ring F *The Thermal Human Body. A Practical Guide to Thermal Imaging*. Singapore: Jenny Stanford Publishing, 2019.

Bergersen TK A search for arteriovenous anastomoses in human skin using ultrasound Doppler. *Acta Physiologica Scandinavica* 1993, 147(2): 195–201.

Biomarkers Definitions Working Group. Biomarkers and surrogate endpoints: Preferred definitions and conceptual framework. *Clinical Pharmacol Ther* 2001, 69: 89–95.

Clark JA Effects of surface emissivity and viewing angle errors in thermography. *Acta Thermographica* 1976, 1: 138–141.

Dzien C, Halder W, Winner H, Lechleitner M Covid-19 screening: Are forehead temperature measurements during cold outdoor temperatures really helpful?. *Wiener Klinische Wochenschrift* 2021, 133(7): 331–335.

Engel J-M, Flesch U, Stüttgen G eds. *Thermological Methods*. Weinheim: VCH, 1985.

Fernández-Cuevas I, Marins JCB, Lastras JA, Carmona PMG, Cano SP, García-Concepción MÁ, Sillero-Quintana M Classification of factors influencing the use of infrared thermography in humans: A review. *Infrared Physics & Technology* 2015, 71: 28–55.

Filippini C, Cardone D, Perpetuini D, Chiarelli AM, Gualdi G, Amerio P, Merla A Convolutional neural networks for differential diagnosis of Raynaud's phenomenon based on hands thermal patterns. *Applied Sciences* 2021, 11(8): 3614.

Fisher G, Foster EB, Pascoe DD Equilibration period following exposure to hot or cold conditions when using infrared thermography. *Thermology International* 2008, 18(3): 95–100.

Getson P, Schwartz RG, Brioschi M, Pittman J, Rind B, Crawford J, Usuki H, Amalu W, Head J Guidelines for breast thermography. *Pan American Journal of Medical Thermology* 2015, 2(1): 26–34.

Goodman PH, Murphy MG, Siltanen GL, Kelley MP, Rucker L Normal temperature asymmetry of the back and extremities by computer-assisted infrared imaging. *Thermology* 1988, 1: 194–202.

Herrick AL, Murray A The role of capillaroscopy and thermography in the assessment and management of Raynaud's phenomenon. *Autoimmunity Reviews* 2018, 17(5): 465–472

Imeri L Thermoregulation as a non-unified system: A difficult to teach concept. *Temperature* 2017, 4(1): 1–8. doi: 10.1080/23328940.2017.1281872

Khallaf A., Williams RW, Ring EFJ, Elvins DM Thermographic study of heat loss from the face. *Thermologie Österreich* 1994, 4(2): 49–54.

Kish MA Guide to development of practice guidelines. *Clinical Infectious Diseases* 2001, 32(6): 851–854.

Kolacz S, Moderhak M, Jankau J New perspective on the in vivo use of cold stress dynamic thermography in integumental reconstruction with the use of skin-muscle flaps. *Journal of Surgical Research* 2017, 212: 68–76.

Love TJ Heat transfer considerations in the design of a thermology clinic. *Thermology* 1985, 1: 88–91.

Madhu H, Kakileti ST, Venkataramani K, Jabbireddy S. Extraction of medically interpretable features for classification of malignancy in breast thermography. In *38th Annual International Conference of the IEEE Engineering in Medicine and Biology Society (EMBC)*, IEEE 2016, pp. 1062–1065.

Marins JCB, Moreira DG, Cano SP, Sillero Q, Soares DD, Fernandes ADA, de Silva CS, Costa CMA, de Amorim PR, Time required to stabilize thermographic images at rest. *Infrared Physics & Technology* 2014, 65: 30–35.

Marjanovic EJ, Britton J, Howell KJ, Murray AK Quality assurance for a multi-centre thermography study. *Thermology International* 2018, 28(1): 7–13.

Melnizky P, Ammer K, Rathkolb O Thermographic findings of the lower extremity in Patients with Type II diabetes. *Thermology International* 2002, 12(3): 107–114.

Midttun M, Sejrsen P Blood flow rate in arterio-venous anastomoses and capillaries in thumb, first toe, ear lobe, and nose. *Clinical Physiology* 1996, 44(16): 275–289.

Moreira DG, Costello JT, Brito CJ, Adamczyk JG, Ammer K, Bach AJE, Costa CMA, Eglin C, Fernandes AA, Fernández-Cuevas I, Ferreira JJA, Formenti D, Fournet D, Havenith G, Howell K, Jung A, Kenny GP, Kolosovas-Machuca ES, Maley MJ, Merla A, Pascoe D, Priego-Quesada JI, Schwartz RG, Seixas ARD, Selfe J, Vainer BG, Sillero-Quintan M Thermographic imaging in sports and exercise medicine: A Delphi study and consensus statement on the measurement of human skin temperature. *Journal of Thermal Biology* 2017, 69: 155–162.

Nugent PW, Shaw JA, Pust NJ Correcting for focal-plane-array temperature dependence in microbolometer infrared cameras lacking thermal stabilization. *Optical Engineering* 2013, 52(6): 061304.

Oliver B, Munro AG, Herrington LC The effect of distance and angle of a smartphone-compatible infrared thermal imaging camera on skin temperature at the midportion of the Achilles tendon. *Thermology International* 2020, 30(2): 51–57.

Pascoe DD, Fisher G Comparison of measuring sites for the assessment of body temperature. *Thermology International* 2009, 19(2): 35–42; Pascoe DD Comparison of measuring sites for the assessment of body temperature: Final amendment. *Thermology International* 2010, 20(1): 36–38.

Pauling JD, Flower V, Shipley JA, Harris ND, McHugh NJ Influence of the cold challenge on the discriminatory capacity of the digital distal–dorsal difference in the thermographic assessment of Raynaud's phenomenon. *Microvascular Research* 2011, 82(3): 364–368.

Plassmann P, Jones CD Recording thermal images and software development for standardisation of thermal imaging. In Wiecek B (ed.) *Thermography and Lasers in Medicine*. Lodz: Akademickie Centrum Grafizno-Marketingowe Lodart S.A, 2003, pp. 19–24.

Plassmann P, Kluwe B Plantar foot assessment using liquid crystal thermography. *Thermology International* 2021, 31(3): 99–100.

Plassmann P, Ring EFJ, Jones CD Quality assurance of thermal imaging systems in medicine. *Thermology International* 2006, 16: 10–15.

Ring EFJ A thermographic index for the assessment of ischemia. *Acta Thermographica* 1980, 5: 35–38.

Ring EFJ Quality control in Infrared thermography. In: Ring E.F.J. and Phillips B., editors. *Recent Advances in Medical Thermology*. New York: Plenum Press, 1984, pp. 185–194.

Ring EFJ, Aarts N, Black CM, Boesiger P et al. Raynaud's phenomenon: Assessment by thermography. *Thermology* 1988, 3: 69–73.

Ring EFJ, Ammer K The technique of infrared imaging in medicine. *Thermology International* 2000, 10(1): 7–14.

Romanovsky AA Thermoregulation: some concepts have changed. Functional architecture of the thermoregulatory system. *The American Journal of Physiology-Regulatory, Integrative and Comparative Physiology* 2007, 292: R37–R46, First published September 28, 2006; doi:10.1152/ajpregu.00668.2006

Rost A *Lehrbuch der Regulationsthermographie*. Stuttgart: Hippokrates Verlag, 1993.

Schwartz RG, Elliott R, Goldberg GS, Govindan S, Conwell T, Hoekstra PP Practice Guidelines Committee of the American Academy of Thermology: Guidelines for neuromusculoskeletal thermography. *Thermology International* 2006, 16: 5–9.

Schwartz RG, Getson P, O'Young B, Campbel J, Brioschi M, Usuki H, Crawford J, Steed P, Ammer K, Serbu G Guidelines for dental-oral and systemic health infrared thermography. *Pan American Journal of Medical Thermography* 2015, 2(1): 44–53.

Schwartz RG, Horner C, Kane R, Getson P, Brioschi M, Pittman J, Rind B, Campbell J, Ehle E, Mustovoy A, Crawford J, Tokman A Guidelines for Breast Thermology n.d. https://aathermology.org/organization-2/guidelines/guidelines-for-breast-thermology/ accessed 19.07.2021

Simpson R, Machin G, McEvoy H, Rusby R Traceability and calibration in temperature measurement: A clinical necessity. *Journal of Medical Engineering & Technology* 2006, 30: 212–217.

Stratford PW, Riddle DL Assessing sensitivity to change: Choosing the appropriate change coefficient. *Health and Quality of Life Outcomes* 2005, 3(1): 23.

Turner TA, Marcella K, Riegel R, Godbold J, Muniz Honório K, Schwartz RG, Campbell J, Brioschi M Veterinary guidelines for infrared thermography. *American Academy of Thermology*, 2019. https://aathermology.org/organization-2/guidelines/veterinary-guidelines-for-infrared-thermography/ accessed 19.07.2021

Usamentiaga R, Garcia DF, Ibarra-Castanedo C, Maldague X. Highly accurate geometric calibration for infrared cameras using inexpensive calibration targets. *Measurement* 2017, 112: 105–116.

Vardasca R The influence of angles and distance on assessing inner canthi of the eye skin temperature. *Thermology International* 2017, 27(4): 130–135.

Wallace JD, Cade CM *Clinical Thermography*. Cleveland: CRC Press, 1975.

Watmough DJ, Fowler PW, Oliver R The thermal scanning of a curved isothermal surface. *Physics in Medicine & Biology* 1970, 15: 1–8.

Westermann S, Stanek C, Schramel JP, Ion A, Buchner HHF. The effect of airflow on thermographically determined temperature of the distal forelimb of the horse. *Equine Veterinary Journal* 2013, 45(5): 637–641.

Weum S, Mercer JB, de Weerd L. Evaluation of dynamic infrared thermography as an alternative to CT angiography for perforator mapping in breast reconstruction: A clinical study. *BMC Medical Imaging* 2016, 16: 43.

Wiecek B, Jung A, Zuber J. Emissivity - Bottleneck and challenge for thermography. *Thermology International* 2000, 10(1): 15–19.

Wilkinson JD, Leggett SA, Marjanovic EJ, Moore TL, Allen J, Anderson ME et al. A multicenter study of the validity and reliability of responses to hand cold challenge as measured by laser speckle contrast imaging and thermography: Outcome measures for systemic sclerosis–related Raynaud's phenomenon. *Arthritis & Rheumatology* 2018, 70(6): 903–991.

Woodrough RE *Medical Infrared-Thermography. Principles and Practice*. New York: Cambridge University Press, 1982.

Zhang HY, Youk T, Lee HK, Song HJ, Yang KH Reference standard temperature data of normal Korean extremities. *Thermology International* 2019, 29(2): 67–72.

3 Basic Approaches of Artificial Intelligence and Machine Learning in Thermal Image Processing

3.1 IMAGE SOURCE

Thermal imaging involves capturing infrared radiations emanating from the source by an array of IR sensors. These sensors are made up of photosensitive materials that convert the incident radiations into electrical signals. Then, these electrical signals are sampled and quantized to produce a thermal image. Advanced thermal imagers consist of thermal detectors called microbolometers which convert the radiant IR energy to a change in resistance which in turn is converted to a change in the electrical signal. An array of microbolometers forms the thermal imaging sensor (Focal plane array IR sensor) with each unit size varying from 0.9 to 14 μm, thus leading to IR resolution ranging from 640×512 to 320×240, respectively, based on the thermal imager type. As the number of pixels increases, the resolution of the camera also increases, producing high spatial resolution images. For example, a 640×512 pixel array thermal camera consists of 327680 measurement points for acquiring the temperature data of the object under focus. Similarly, a 320×240 pixel resolution thermal camera provides 76800 temperature measurement points of the scene (Vollmer and Möllmann, n.d.).

Figure 3.1 shows a few sample images of a human face and a laboratory room, respectively, imaged by a thermal imaging camera. The images are illustrated in Rainbow color mode, a default color mode in various FLIR cameras to show the temperature distribution of the scene. The pseudocolor is provided with reference to a color bar showing the range of temperature values in the given thermal image.

3.2 DATA TRANSFER FROM CAMERA TO THE ANALYZING SOFTWARE

Thermal scanners or imagers generally provide the thermal images as a stream of digital data (16-bit radiometric) which could be accessed by the proprietary software programs and could also be acquired using computation tools such as Matlab.

DOI: 10.1201/9781003245780-3

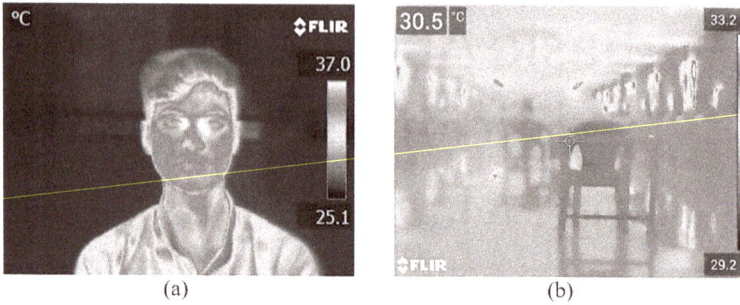

(a)   (b)

FIGURE 3.1 Thermal image taken by a FLIR camera with a resolution of 320×240: (a) A human face and (b) a room.

Modern thermal cameras have streaming capability which usually provides network interface links such as GigeCam, which is an industry standard for high-speed data transfer through Ethernet port or by USB video class (UVC) port. In addition, most thermal cameras have the ability to store the thermal images in mass storage devices, such as micro-SD memory cards, provided in the equipment itself. The stored data from the memory card can be transferred to a PC using a serial port communication for further processing (Teledyne FLIR, n.d.).

3.3 PRE-PROCESSING

Thermal images are similar to visible images; however, each pixel is a function of temperature values compared to visible images that represent the intensity values of the scene. Image processing tasks that are widely applied for visible images, such as image enhancement, edge detection, feature extraction, image segmentation, and object detection, are also used for thermal images. In this section, we review some of the image processing tasks used for image analysis.

However, there is an important difference between the gray value of a thermal image and the intensity value of other gray shade images. Temperature is an interval scaled quantity because the centigrade scale lacks an absolute zero, thus, temperature pixels can neither be multiplied nor be divided by themselves or raised to power. While image processing improves image features such as resolution, contrast, and sharpness, it may severely affect temperature measurements due to procedures not allowed for interval scales.

The foundations in mathematics of image processing include linear algebra, probability and statistics, calculus, and formulation of machine-learning algorithms (Pattanayak, 2017). Linear algebra deals with linear equations and their representation by vectors and matrices. Differential calculus focuses on how quickly physical quantities change, e.g., how much intensity of an image varies with respect to spatial domain, while integral calculus provides means to calculate the sum or aggregation of a quantity, e.g., summation of intensity values over the defined area in the image. Probability describes the rate of outcomes of an experiment in the sample space and statistics allow analyzing the distribution of outcomes in experiments with respect to their occurrence by chance.

3.3.1 THERMAL IMAGE COLOR MODES

A thermal imaging system uses sensors to quantity the IR radiations emanating from an object and each point in a thermal image represents a temperature reading. Thus, a thermal image in its original format provides the radiometric (temperature) data for each pixel. Often it is required to add pseudocolor to differentiate the temperature of different objects in the thermal image. Hence, the manufacturer of the thermal camera provides provision to enable different color modes for the temperature data in the thermal image by means of software tools either in the equipment itself or as a standalone application. Figure 3.2 provides the different color modes typically available in a FLIR camera (Model: E54) for showing the different temperature regions in the scene. The pseudocolors are referenced with a temperature color bar to enable the user to determine the temperature value of various objects in the thermal image. Figure 3.2 shows a thermal image of

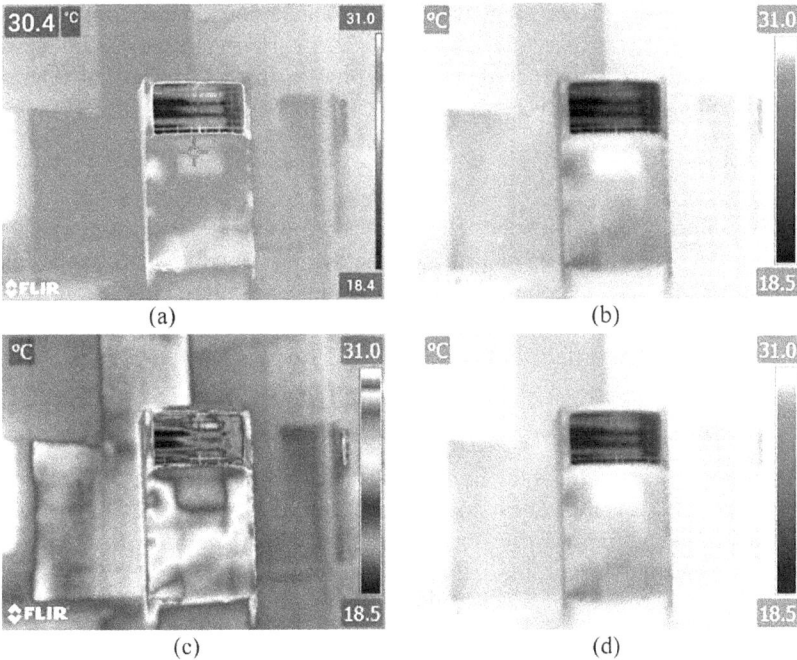

FIGURE 3.2   Representation of thermal images in different color modes widely available in commercial thermal imaging system: (a) Rainbow, (b) Grayscale, (c) Rainbow High Contrast, and (d) Ironbow.

an air-conditioning system in different color modes such as (a) Rainbow, (b) Grayscale, (c) Rainbow High Contrast, and (d) Ironbow, respectively. Rainbow and Grayscale color modes are widely used owing to consistency in the representation of temperature data when displayed on a monitor or for a digital printout.

3.3.2 IMAGE ENHANCEMENT

Thermal images are often used for various studies such as detection of hot or cold objects, human crowd flow monitoring, video surveillance, face detection, fever screening, and in various other fields. The thermal images obtained from the imaging system need to be pre-processed before applying for complex image processing tasks such as object detection and recognition. Hence, common image pre-processing steps such as image smoothening, sharpening, filtering, and contrast enhancement are often done to enhance the thermal images. In this section, we review some of the image processing tasks used widely in thermal image processing.

3.3.2.1   Image Smoothening and Sharpening

Thermal images are often degraded by various factors such as ambient environment, sensor noise, optical lens blurring, and transmission noise. Thermograms corrupted by random noise often manifest as granular noise and restrict the usage of the thermograms in object detection applications. Hence, the noises are filtered either by spatial or frequency domain filtering. In the case of spatial filtering, a smoothening kernel such as the average filter mask which could be defined

as $h[m,n] = \dfrac{1}{M} \sum\limits_{(k,l)\in N_i} f[k,l]$ is convolved with the image corrupted by random

noise specifically Gaussian noise. Here, "h" is the averaging or smoothening kernel and "M" is the number of pixels in the neighborhood "N_i." k, l represents the number of rows and columns of the input image and m, n indicates the pixel coordinates of the smoothening kernel. In the case of frequency-based noise filtering, the input image is transformed to the Fourier domain, and a frequency mask

which could be defined as $H(u,v) = \begin{cases} 1 & \text{if } FI(u,v) \leq D_0 \\ 0 & \text{if } FI(u,v) > D_0 \end{cases}$ is multiplied with the

Fourier image. Here, "H" is the frequency mask, "FI" is the center of the Fourier Image mask, "D_0" is the cut-off spatial frequency, and (u, v) represents the pixel coordinates in the transformed domain. The noise is filtered and the processed image is converted back to the spatial domain by performing an inverse Fourier transform.

In certain thermograms, due to improper lens focus, the images lack sharp details such as the edges being blurred. To sharpen the images and enhance the edges in the input thermograms, image sharpening is usually done. One of the commonly used image sharpening kernels that is based on gradient (directional change with respect to intensity or color in the image) computation is the Laplacian function, which is defined as $L(x,y) = \dfrac{\partial^2 I}{\partial x^2} + \dfrac{\partial^2 I}{\partial y^2}$. Here, L is the Laplacian Operator, and "I" is the input image, x and y represent the pixels with respect to the horizontal direction and the vertical direction, respectively. In the case of frequency-based image sharpening also related to high pass filtering, the input image is transformed into a Fourier domain and a frequency mask which could be defined as $H(u,v) = \begin{cases} 1 & \text{if } FI(u,v) \geq D_0 \\ 0 & \text{if } FI(u,v) < D_0 \end{cases}$ is multiplied with the Fourier image. Here, "H" is the frequency mask, "FI" is the center of the Fourier image mask and "D_0" is the cut-off spatial frequency. The low-frequency component is filtered and the sharpened (edge enhanced) processed image is converted back to the spatial domain by performing inverse Fourier transform (Gonzalez & Woods, 2018).

3.3.2.2 Histogram Equalization

Thermal images show the temperature distribution of a particular scene and are widely displayed in different color modes. Often the contrast of the images is compromised due to environmental factors. Thus, it is necessary to improve the contrast of the thermograms. Histogram equalization is an efficient contrast-enhancing technique widely used in many image enhancing applications.

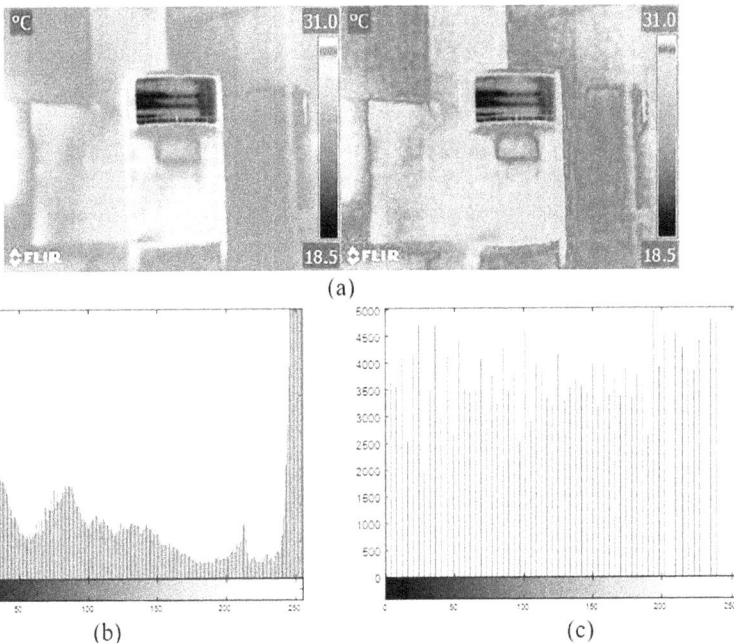

FIGURE 3.3 Histogram equalization of a thermal image: (a) Input image (left) and enhanced image (right), (b) histogram of the input image, and (c) histogram equalized thermal image.

Figure 3.3a (left image) shows the thermogram of an air-conditioning system and its histogram is shown in Figure 3.3b).

As the input image is in RGB color space, the application of histogram directly on the color image leads to distorted images. Hence, the image is changed to HSI color space and thus it is separated into color and intensity images. Histogram equalization is then performed on the intensity data. Observing the histogram of the input thermogram, we could notice that the number of pixels for each gray level is not evenly distributed. Therefore, the contrast of the image is reduced. Hence, Histogram equalization could be used to balance the number of pixels in each gray level. Figure 3.3c shows the equalized histogram plot where the gray level is equally distributed and Figure 3.3a (right image) shows the enhanced thermal image after histogram equalization. By performing histogram equalization, we enhance the contrast of the image for better visual perception. This doesn't affect the accuracy of temperature readings.

3.3.3 MULTISPECTRAL IMAGES

Thermal images provide information based on IR radiations emanating from the source. The sensors are configured to detect these IR radiations in a particular frequency band. Thermal cameras which are sensitive to different IR bands are available based on the application or the object/scene of interest.

The IR bands available in modern thermal cameras are categorized as (a) Short Wave IR (SWIR) which ranges from 0.9 to 1.7 μm, (b) Medium Wave IR (MWIR) band which ranges from 1 to 5 μm, and (c) Long Wave Infrared (LWIR) band which ranges from 8 to 15 μm. Often, the entry-level thermal imaging system uses IR radiations in a particular band, namely in the LWIR range. However, modern and high-end advanced thermal imaging systems are capable of sensing the IR radiations in the three bands. Hence, the images obtained from these IR spectral bands are termed multispectral thermal images. In addition, it is also customary to use a visible camera to capture the scene based on radiations falling in the visible spectrum of electromagnetic radiations. An example of an electrical system and a face imaged using a visible camera and a thermal camera is shown in Figure 3.4. Often, these multispectral images are fused to gain more knowledge about the scene or the object for post processing such as object extraction and recognition as they provide complementary information about the scene. However, when fusing images such as thermal or visible images from different sources, it requires the images to be in uniform geometric space for proper post processing in image applications.

3.3.4 IMAGE REGISTRATION

Image registration is an essential image processing step when information from different imaging modalities (multimodal) such as thermal and visible images are fused for disease diagnosis and detection analysis. However, the registration

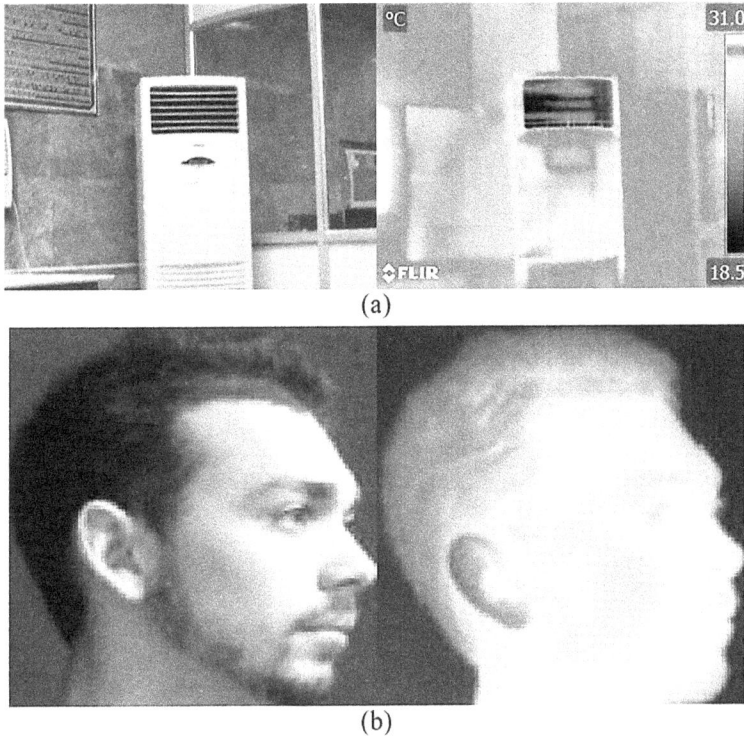

FIGURE 3.4 Multispectral images showing visible and thermal images of (a) air-conditioning system and. (b) human face. (See Panetta et al., 2020.)

of thermal and visible images is not straightforward as the image sensor is constructed discretely and the field of view of the sensors is also different based on the position of the two cameras. An example of two image sensors capturing a scene is provided in Figure 3.4. There is a slight misalignment between the two images in Figure 3.4b – that is the geometric axis is not uniform for both the images. This could be seen in Figure 3.5. Hence, it is suggested to align or register the images obtained from different modalities for proper image fusion. There are a plethora of image registration schemes available in the literature which could be commonly categorized as intensity-based image registration and feature point-based image registration techniques. In the case of intensity-based registration methods, one of the input images (moving image–thermal image) is transformed geometrically based on affine transformation with respect to the fixed image (visible image). The optimal parameters of the affine transformation such as the rotational angle, translational distance, and shearing parameters are obtained by maximizing an intensity-based similarity parameter such as the mutual information between the two images (thermal and visible

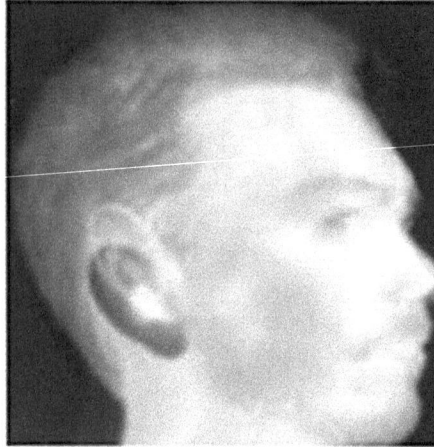

FIGURE 3.5   Illustration of the need for image registration for thermal and visible images for proper image fusion. We observe that the two images are slightly misaligned because of the position of the two cameras.

images). In the case of the feature point selection method, feature point extraction methods such as SIFT, SURF, or ORB are used to determine unique feature points between the two input images. Then, based on the similarity of the feature points in the two images, the points are then matched, and the image is transformed by a projective homography (Wu et al., 2018). Homography refers to any two images with the same planar space that are connected by means of applying the transformation of a point in one image space into corresponding points in another image.

The moving image is transformed to match the fixed image in spatial and geometric space thus leading to image registration. Once the input images obtained from different image sensing systems are registered, they could be used for image fusion (Ma, Ma, and Li, 2019).

3.3.5 IMAGE FUSION

Multispectral images that consist of thermal images from different IR bands or visible images often contain complementary information that could be combined to achieve better human perception and enhance post-processing tasks in image analysis such as feature extraction, segmentation, and object detection and recognition. For example, in the application of face detection for crowd monitoring and security surveillance, thermal and visible images are often obtained from different sensor systems. The accuracy of face detection and recognition is improved when the two images are combined by some fusion technique. There are numerous works related to image fusion, however, methods based on multiresolution wavelet decomposition are reported with good fusion performance (Pajares and

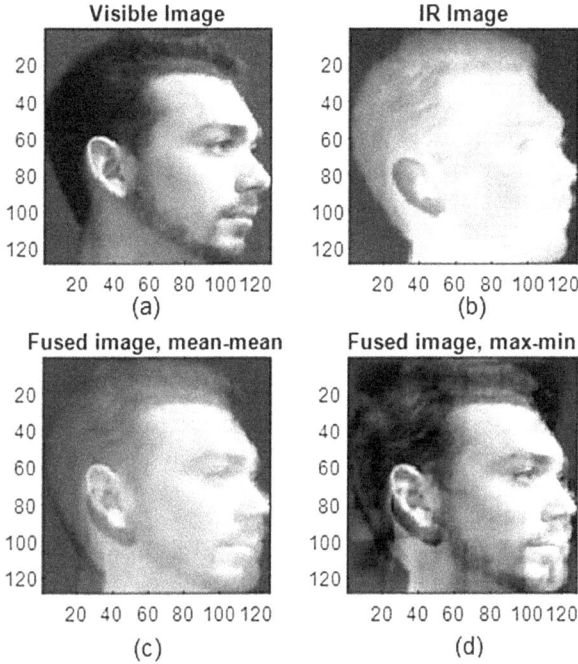

FIGURE 3.6 (a) Thermal and (b) visible image fusion based on wavelet decomposition. (c) Mean of wavelet coefficients and (d) max–min of wavelet coefficients.

de la Cruz, 2004). Figure 3.6 shows the application of discrete wavelet transform for the fusion of thermal and visible images for face analysis.

The wavelet method of image fusion involves decomposing the thermal and visible images to a pre-defined level such as 2, 3, or 4. Once, the required level of decomposition is selected, the wavelet coefficients from the thermal and visible images are combined to get an enhanced processed image (Sappa et al., 2016). The final fused image is obtained by taking the inverse discrete wavelet transform. Figures 3.6c and 3.6d show the fused image by using mean–mean and max–min fusion techniques for fusion approximation and detail coefficient, respectively (Ma et al., 2020; John et al., 2017).

3.4   SEGMENTATION TECHNIQUES

Medical image segmentation plays a pivotal role in computer-aided diagnosis in physiological systems. Segmentation is considered the most essential step in image processing as it separates the region of interest (ROI) from the background using an automatic or semiautomatic process. Medical image segmentation is used to partition the image into important regions such as an organ or tissue, or to detect anomalies in the image such as tumors, cancer, lesions, and inflammations. Although the segmentation algorithm that needs to be used differs according to the application, there is a general criterion that should be fulfilled by the algorithm. A good image segmentation algorithm must fulfill the following criteria:

(i) It should be replicable, i.e., when used on the same data; the results should not keep changing every time.
(ii) It should be consistent in detecting the ROI even in different datasets.
(iii) The algorithm should be accurate in detecting the ROI in the most precise way. Accurate results can be achieved only when the segmentation is accurate.
(iv) The algorithm should be time-efficient because large computing times are not favorable for real-time applications.

Separation of the ROI from the background may be accomplished using several different automated or semi-automated techniques, depending on the application and also the nature of the input image. Nowadays, the more commonly used method is the clustering-based technique. There are a broad variety of clustering techniques to choose from, including k-means clustering, divisive clustering, mean shift clustering, hierarchical clustering, fuzzy c-means clustering (FCM), and neutrosophic c-means clustering. Segmentation may also be done based on the thresholding and histogram analysis, edge detection, graph partitioning, and watershed-based method (Saparudin et al., 2018). Also, recent advances in technology have now enabled us to make use of deep learning models such as u-net for image segmentation.

3.4.1 CLUSTERING-BASED SEGMENTATION ALGORITHMS

Cluster-based image segmentation involves the grouping of the image pixels into groups also known as clusters (Mittal et al., 2021). The pixels in each group are as similar to each other as possible, while pixels belonging to different groups are as dissimilar to each other as possible. Also, the distance between cluster centers is also kept as far as possible. The common principle behind the different clustering algorithms is that initially, the number of clusters is to be specified by the user. Next, the centroid is calculated, and the pixels are allotted to each cluster accordingly. More iterations are performed to the centroid calculation until it no longer changes, i.e., the algorithm converges. Some of the important and frequently used algorithms in medical image processing are explained as follows.

3.4.1.1 k-Means Clustering

k-Means clustering algorithm is perhaps one of the most extensively used cluster-based algorithms in recent times, due to its ease of computation as well as versatility for different applications. k-means clustering is an iterative algorithm that divides the data set into k cluster numbers, with each data point being a member of only one group. The aim of this algorithm is to make the data points as identical as possible within each cluster, at the same time keeping the clusters as far away from each other as possible. Within each cluster, the sum of the squared distances between each data point as well as the centroid of the cluster is kept as minimum as possible. This means that the data points within the cluster are very similar (homogenous) to each other. The algorithm is elaborated in steps as follows:

(i) Allocate the number of clusters (k).
(ii) Define the initial centroids by randomly choosing k data points as the centroids (without repetition).
(iii) Calculate the squared distance sum between data points in each cluster and the centroids of all the clusters.
(iv) Allocate every data point to the nearest cluster.
(v) Update the centroids for the clusters by computing the mean value of all the data points in each cluster.
(vi) Continue the iterations until there is no change to the centroids

If the Euclidean distance is not used, the algorithm does not converge. In order to use other distance measures, a few modifications of the k-means algorithm have been proposed such as the k-medoids and spherical k-means algorithm.

3.4.1.2 Hierarchical Clustering

Hierarchical Clustering, also known as AGNES (agglomerative nesting), is a bottom-up clustering approach. The Data points present inside each cluster are as proximate to each other as possible, and data points from different clusters are as distant from each other. In hierarchical clustering, pairs of clusters are merged as they move up the hierarchy. The algorithm can be explained as follows:

(i) Each data point or pixel is made as a single point cluster. At this point, we have N number of clusters.
(ii) Two data points closest to each other are taken and merged to form one cluster. Now, we have N − 1 clusters.
(iii) Again, take any two close single point clusters and merge them to form a bigger cluster. This gives us N − 2 clusters.
(iv) The above steps are repeated until there is only one big cluster.

The hierarchical clustering algorithm is often represented in a graph that is known as a dendrogram, which can either be a column or a row graph. It represents the hierarchical relationships between different sets of data. By looking at a dendrogram, it is possible to understand how each cluster was formed.

3.4.1.3 Divisive Clustering

Divisive clustering algorithm, otherwise known as DIANA (divisive analysis clustering), is a top-down clustering approach where all the pixels are assigned to a single cluster and then partitioned into smaller clusters during each iteration.

Finally, we are left with N number of clusters, with only one data point in each cluster. This method is the exact opposite of the hierarchical clustering method. The steps involved may be explained as follows:

(i) Initially, all the pixels in the image belong to one data cluster.
(ii) Next, the cluster is partitioned into two similar clusters at least.
(iii) The above step is repeated until the desired number of clusters has been reached.

3.4.1.4   Density-Based Clustering

The previously discussed clustering methods are most suitable for well-separated and compact clusters and do not perform efficiently when there is noise and outliers present in the data. Density-based clustering (also known as DBSCAN) is an unsupervised clustering algorithm that determines the clusters of arbitrary shapes and the noise in a spatial database (Ester et al., 1996). The DBSCAN algorithm requires two parameters:

(i) **Epsilon:** It described the neighborhood of each pixel. If the distance between two pixels is equal to or lesser than the epsilon value, then they are considered neighbors. Otherwise, they are not considered neighbors.
(ii) **Minimum Points:** Defines the minimum number of neighbors within the epsilon radius. If the dataset is larger, a larger value of minimum points must be chosen.

The DBSCAN algorithm first starts by finding all the neighboring pixels within the epsilon radius for each data point, and core points are identified (Sander et al., 1998). If each core point has not already been assigned a cluster, a new cluster is assigned. Each density-connected point is found and ascribed to the same cluster as the core point. The iterations are repeated for each unvisited data point in the image. At the end of the algorithm, the pixels that have not been assigned a cluster are the noise.

3.4.1.5   Fuzzy C-Means Clustering

FCM is an unsupervised-based clustering algorithm. It has been applied to a wide range of applications such as feature extraction, feature analysis, target recognition, and design of classifiers. Fuzzy segmentation is more favorable than crisp segmentation because data points can belong to more than one cluster, leading to better performance. It is also found to give better results than the conventional k-means segmentation algorithm. The FCM algorithm is as follows:

(i) Randomly select "c" number of clusters.
(ii) Compute the fuzzy membership value (μ_{xy}).
(iii) Compute fuzzy centers of the clusters (V_y).
(iv) Perform repeated iterations until the value of the objective function "Y" is minimized.

The objective of this algorithm is to minimize the function:

$$Y(u,v) = \sum_{x=1}^{n}\sum_{y=1}^{c} (\mu_{xy})^{m} \left\| X_x - V_y \right\|^{2} \tag{3.1}$$

where X_x represents the data point *and* V_y indicates the cluster center, $\|X_x - V_y\|^2$ represents the Euclidean distance between the xth data and yth cluster center, $\mu_{xy} =$ Fuzzy membership of the xth data to the yth cluster, n = number of data points, m = fuzziness index ranging from 0 to 1.

3.4.1.6 Neutrosophic C-Means Clustering

It is a powerful general framework that was proposed in recent times and has since been used for several applications. Neutrosophic set theory has been previously used in image processing for applications such as de-noising, segmentation, and thresholding (Guo and Sengur, 2015).

For each pixel, the neutrosophic c-means algorithm calculates the membership of determinate and indefinite clusters at the same time. The membership "T" is the membership of the pixel to the determinate cluster, while "I" and "F" are used to describe membership to the two types of indeterminate clusters: the ambiguity clusters and the outlier clusters. The outlier cluster helps us to reject the pixels that are too far away from cluster centers. Ambiguity clusters help us to consider the pixels that are situated close to the cluster boundaries. In this algorithm, the membership function is more resistant to noise as compared to other algorithms.

3.4.1.7 Mean Shift Clustering

Mean shift clustering algorithm assigns each pixel in the image to a cluster centroid. The concept of kernel density estimation is used in this algorithm. The direction in which the closest centroid is selected depends upon which direction has the greatest number of points nearby. In each iteration, the pixels will move closer to the direction where the most data points are located, which finally leads to the cluster center. At the end of the algorithm, each pixel in the image will have been assigned to a cluster. The advantage of mean shift clustering is that the number of clusters does not need to be specified beforehand. The number of clusters is determined by the algorithm with respect to the data.

3.4.2 THRESHOLD-BASED SEGMENTATION

Thresholding is perhaps the basic method of image segmentation, generally applied to grayscale images. This method is suitable for images where there is an object or region of uniform brightness that is placed against a dark background. In such cases, a threshold can be applied to segment the object from the background. If thresholding is applied to a grayscale image, the resulting image will be a binary image that consists of pixels that can have one of exactly two colors, usually black and white.

The thresholding method is where each pixel in an image will be replaced with a black pixel if the intensity value of the pixel is less than the specified threshold value. If the intensity value is greater than or equal to the threshold value, then the pixel will be replaced with a white pixel. Mathematically, the threshold can be defined as follows:

$$Y_{i,j} = \begin{cases} 1, if\ X_{i,j} \geq T \\ 0, if\ X_{i,j} < T \end{cases} \qquad (3.2)$$

where $Y_{i,j}$ is the output pixel value in the coordinate (i, j), $X_{i,j}$ is the coordinate of the pixel in the input image, and T is the threshold value.

The selection of the threshold value is therefore very important in the segmentation of the image. The selection of threshold may be done by analyzing the histogram of the image. Completely automated thresholding is also possible, through techniques such as clustering, spatial analysis, and entropy-based methods.

It is possible to apply thresholding on color images as well. This process is known as multiband thresholding, as red, green, and blue (RGB) images have three different components, i.e., red, green, and blue. It is possible to separate the three components – apply the threshold for each separately, and then join the components back together by using the AND operation. Hue, saturation, and vertical intensity (HSV) and hue, saturation, and lightness (HSL) color models are commonly used in multiband thresholding, although it is possible to use RGB and cyan, magenta, yellow, and black (CMYK) models as well.

3.4.3   REGION-BASED SEGMENTATION ALGORITHMS

In region-based segmentation, the underlying assumption is that objects are separated by perceptual boundaries such as neighborhood features or different textures. The region-based segmentation process is based on the continuities that are present in the image. The main objective behind region-based segmentation is to partition or divide the image into different regions, depending on the specified conditions. The criteria that should be fulfilled by a good region-based segmentation algorithm are described as follows:

  (i) Every pixel in the image should belong to a region.
 (ii) The pixels in each region should be connected in some predefined sense.
(iii) Each region should be disjoint from the other.

The different types of region-based segmentation algorithms that are commonly implemented in medical image segmentation are region-growing, region-merging, and active contour segmentation. They may be either a bottom-up or a top-down approach. Both approaches have been discussed in the subsequent sections.

3.4.3.1 Region Growing

Region-growing is a bottom-up region-based segmentation technique. In this algorithm, a set of "seed points" are first initialized during the start of the algorithm (Erwin, 2020). The selection of these seed points depends upon the application. The region-growing algorithm may be summarized as follows.

(i) Initially, a set of small areas in the image are merged iteratively, according to similarities between them.
(ii) Next, a seed point is selected and compared with the neighboring pixels.
(iii) The region is grown from the seed pixel, by adding in surrounding pixels that are similar to the seed pixel.
(iv) The growth of the region stops when there are no longer any surrounding pixels that are similar to the seed point up to a certain degree.
(v) When the growth of a region has stopped, a new seed point is selected that does not already belong to a grown region, and the process is repeated.
(vi) The algorithm ends when all the pixels in the image belong to a grown region.

The choice of seed points often influences the segmentation results. Therefore, care should be taken to properly select the seed points, either by histogram analysis or some other relevant method. To overcome some of the undesirable effects of region-growing such as the domination of one region over others or unresolved ambiguities in the region, a technique known as simultaneous region-growing has been developed. In this method, all the regions are grown simultaneously, and no region is allowed to dominate over other regions. Also, similarities of neighboring regions are considered, and very similar regions will gradually expand into each other.

3.4.3.2 Region Merging

Region merging starts with a uniform seed region. Once a uniform seed region has been fixed, the neighboring pixels are then merged with the seed region until the neighboring pixels do not conform to the uniformity criteria, i.e., the neighboring pixels are dissimilar to the seed region. Then, the region is extracted from the image, and a different seed region is used to merge other regions of the image. A disadvantage of using this approach is the large computing time that is necessary if you need to extract minute segmentation details from the image. Also, similar to region-growing, the final segmentation result will vary depending on the seed region which has been chosen.

3.4.4 Active Contour

Active contour models are based on curve flow to extract the exact ROI from the image (Hemalatha et al., 2018). Contours are essentially the boundaries that surround ROIs. The collection of points that are present in the contour undergoes an interpolation process, which may be either linear, splines, or polynomial. Active contour models are energy-based segmentation models, i.e., they separate the regions of the required intensity values based on energy forces. Here, the term energy represents external constraint forces and image forces that pull it toward features such as lines and edges. There are quite a few types of active contour models such as the traditional snake, balloon, and gradient vector flow model.

The snake model basically uses the internal and external forces exerted by the contour curve in order to segment the image. The balloon model is similar to the snake model, except there is an additional inflation force that is used to define the curvature. The gradient vectors of the flow of the curve are used to describe the contour of the model in the gradient vector model.

3.4.5   EDGE DETECTION FOR IMAGE SEGMENTATION

While region-based methods are based on the continuities in the image, edge-based segmentation relies upon the discontinuities in the image. This process classifies sharp discontinuities in the image, i.e., abrupt changes in the intensity value. These discontinuities are considered as the edges or boundaries of regions in the image. Representation of the image as boundaries or edges significantly reduces the amount of data to process, while at the same time retaining all the useful information in the image regarding the shape and orientation of the regions. Edge detection is commonly seen in applications where it is necessary to detect the boundary of a region such as the boundary of a tumor or lesion (Muthukrishnan and Radha, 2011). Some of the most used edge detection techniques are discussed.

The criterion to be followed by edge detection algorithms is as follows:

(i) The edges should be detected with an error rate as low as possible, i.e., the algorithm should identify as many edges from the image as possible.
(ii) The edge point identified should precisely localize on the center of the edge.
(iii) The edge present in the image should only be pointed once.
(iv) False edges should not be produced.

3.4.5.1   Robert Edge Detection

The Robert edge detection algorithm emphasizes regions of high spatial intensity that usually correspond to edges (Sharma and Aggarwal, 2010). It performs a simple 2D gradient operation on the input image. The input image to this operation is a grayscale image, and the output image will be the same size as the input image. The operator is known as Robert's cross operator and consists of 2×2 kernels which are shown in Figure 3.7.

3.4.5.2   Sobel Edge Detection

The Sobel edge detection technique is quite similar to that of Robert's. Once again, a gradient operation is performed upon the input grayscale image resulting in an output image of the same size as the input image. Each point in the output

-1	0	0	-1
0	+1	+1	0
(a)		(b)	

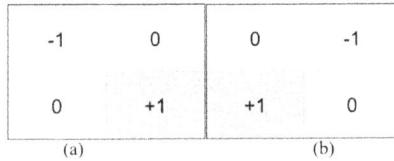

FIGURE 3.7   Robert kernel: (a) Gradient in X-direction and (b) gradient in Y-direction.

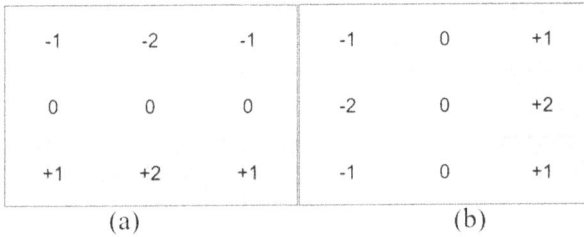

-1	-2	-1	-1	0	+1
0	0	0	-2	0	+2
+1	+2	+1	-1	0	+1
(a)			(b)		

FIGURE 3.8   Sobel kernel: (a) Gradient in X-direction and (b) gradient in Y-direction.

image corresponds to the absolute gradient magnitude of that point in the input image. One difference between the Sobel method and the Robert method is the size of the kernel that is used in the gradient operation. The Sobel method makes use of a 3 × 3 kernel, as shown in Figure 3.8a, where the second kernel shown in Figure 3.8b is obtained by rotating the first kernel by 90°.

3.4.5.3   Prewitt Edge Detection

The Prewitt edge detection is easier to compute than the Sobel method; however, it might sometimes produce noisier results. The Prewitt operator measures the gradient of the image intensity at every point and also gives the largest possible increase from high to low intensity, as well as the rate of change in that direction. The operator used in this method is also of the size 3 × 3 and is calculated in eight possible directions. The direction with the highest response is then selected as the final as shown in Figure 3.9.

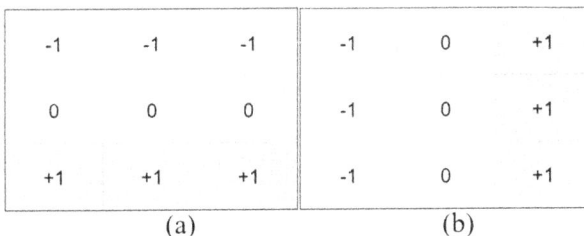

-1	-1	-1	-1	0	+1
0	0	0	-1	0	+1
+1	+1	+1	-1	0	+1
(a)			(b)		

FIGURE 3.9   Prewitt kernel: (a) Gradient in X-direction and (b) gradient in Y-direction.

0	-1	0	-1	-1	-1
-1	4	-1	-1	8	-1
0	-1	0	-1	-1	-1

(a)   (b)

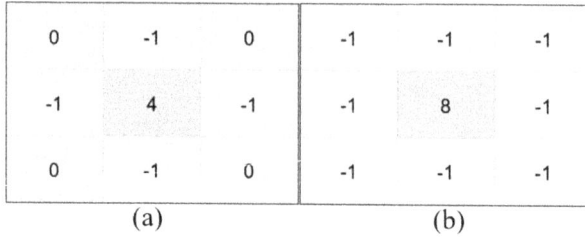

FIGURE 3.10 Laplacian operator for edge detection.

3.4.5.4 Marr–Hildreth Edge Detection

The Marr–Hildreth edge detection algorithm performs convolution on the image with the Laplacian of the Gaussian operator. Then, zero-crossings are detected in the filtered image to obtain the resultant edges. The operator used in this method is sometimes referred to as the Mexican Hat wavelet due to its shape when turned upside down. The digital implementation of the Laplacian is made through the following mask shown in Figure 3.10.

The Marr–Hildreth method suffers from two limitations. Firstly, the algorithm generates several false edges (responses that do not correspond to real edges in the image). Secondly, there is a localization error when there is a curved edge. Nowadays, better detection methods such as the Canny edge detection are available, which do not have such limitations.

3.4.5.5 Canny Edge Detection

The Canny edge detection is one of the most standard techniques in recent times. It outperforms the older algorithms and even some of the newer algorithms that have been developed after it. Due to its superior performance and industry standards, this method is of utmost importance for edge detection and has been discussed in detail as follows.

(i) **Application of Gaussian Filter:** Gaussian filter is applied to the image in order to smooth it and filter out the noise to prevent the detection of false edges. This is carried out by convolution of the image with a Gaussian filter kernel. A Gaussian filter kernel of size $(2k + 1) \times (2k + 1)$ is given by

$$G_{ij} = \frac{1}{2\pi\sigma^2} \exp\left(-\frac{\left(i-(k+1)\right)^2 + \left(j-(k+1)\right)^2}{2\sigma^2}; 1 \leq i, j \leq (2k+1)\right) \quad (3.3)$$

where σ represents the Gaussian kernel width, i, j indicates pixel coordinates of the input image, and k is the number of integers.

(ii) **Computation of Intensity Gradient:** The edges in an image can be pointing toward any direction so the canny edge detection algorithm employs four filters that detect horizontal, vertical, and two diagonal edges in the filtered image.

(iii) **Non-maximum Suppression:** This is a method that is done to make the edges thin. The edge strength of the current pixel value is compared with the edge strength in the positive and negative gradient directions. If the edge strength is maximum compared to other pixels in that mask in the same direction, it is preserved; otherwise, it is suppressed.

(iv) **Double Thresholding:** High and low threshold values are defined to filter out the pixel values with weak edge strengths. This step is done to get a more accurate representation of the real images in the image and gets rid of any extra or false edges, if present.

(v) **Hysteresis:** Hysteresis or blob analysis is performed to analyze the weak edges that are connected to the strong edges, and finally decide whether they should be preserved or not.

3.4.6 Watershed Segmentation

Watershed segmentation is a topography-based, morphological method that is applied to grayscale or gradient images for segmentation purposes. Watershed transform works on the conceptual idea of "immersion." Every local minimum in a grayscale image may be visualized as a catchment basin having a hole in the bottom. Water is then progressively filled through the hole and into the basin, starting from the lowest intensity minima. As the water fills the basins and reaches the top, the water might overflow and touch the water from the adjacent basins. Right at the moment when water from two or more adjacent basins are about to merge, conceptual dams are constructed to prevent the water from merging. At the end of this watershed process, each basin is now enclosed by a dam, which corresponds to the watershed lines in the output image. Thus, watershed transform may be applied to detect edges in the image.

3.4.7 Deep Learning Models for Image Segmentation

Convolutional neural networks (CNNs) are being increasingly implemented in recent times for image segmentation, feature extraction, as well as classification. U-Net is a CNN architecture that was proposed in 2015 by Ronneberger et al. (2015) specifically for the purpose of segmenting biomedical images. U-Net is considered one of the most successful segmentation models in terms of both architectures as well the pixel-based image segmentation. By performing a bit of elastic augmentation, the U-Net model only requires a small number of annotated images to train a large batch of images for segmentation and has shown to produce commendable results.

3.5 ARTIFACT SUPPRESSION

Thermal cameras have a tendency to enhance the visibility of near and distant objects in dark environments. But during the object detection and recognition process, the thermal reflections formed in the image create a kind of artifact. As the electromagnetic radiation emitted by the object gets reflected off the floors and walls, it creates a shadow-like appearance around the object in the image. These shadow-like areas around the object are called thermal reflections or thermal artifacts. The thermal reflections may vary depending on the factors such as the object's temperature, environment surrounding the object, heat generated in the range of reflection, and nature of the material. Hence, to remove or eliminate thermal artifacts, deep learning techniques have been used in recent literature. The Mask-R-CNN is used to identify the thermal reflection areas in the thermal images (Batchuluun et al., 2019). Then, the detected areas are removed by using post-processing methods such as morphological dilation and complements. In addition to Mask-R-CNN, several state-of-the-art methods such as SegNet, and cycle generalized adversarial network (GAN) could be used to detect the thermal reflections or artifacts.

3.6   FEATURE EXTRACTION AND CLASSIFICATION

Feature extraction is an essential step in the machine learning process, wherein useful data and measurements are derived from the set of images that are informative, non-redundant, and help to train the machine learning models with a much better performance. In deep learning models, feature extraction layers are often associated with the reduction of dimensionality. There are various algorithms that may be used to extract useful features from the picture, some of which have been listed below:

- Gray Level Co-occurrence Matrix (GLCM)
- Speeded-Up Robust Features (SURF)
- Gray Level Run Length Matrix (GLRLM)
- Local Binary Pattern (LBP)
- Gray Level Size Zone Matrix (GLSZM)
- Local Binary Gray Level Co-occurrence Matrix (LBGLCM)
- Scale-Invariant Feature Transform (SIFT)
- Local Directional Pattern (LDP)
- Segmentation based Fractal Texture Analysis (SFTA)

3.6.1   GRAY LEVEL CO-OCCURRENCE MATRIX

The GLCM is based on a statistical method that is used for analyzing the texture feature which gives information about the spatial relationship of pixels. In an image, both the column and row will be identical to the number of gray levels denoted as "g" in the GLCM matrix. The texture of the image is used to calculate the pixel with particular values in a specific spatial relationship. Data are divided into first, second, and higher-order statistics. In that way, the second order is a statistical method in the gray level co-occurrence matrix.

GLCM has many statistical parameters such as entropy, energy, homogeneity, correlation, and mean. Each parameter gives different information about the images. The parameters are explained as follows (Mall et al., 2019):

(i) Autocorrelation calculates the value of the quality and roughness of the texture in the image.

(ii) The energy difference moment is a measurement of the image's homogeneous patterns. Energy is also known as the second moment of angular momentum or homogeneity.

(iii) The cluster prominence calculates the skewness and asymmetry of the GLCM matrix. The larger value intimates many asymmetries about the average while the lowest values represent a top close to the average values and low variation about the mean.

(iv) The cluster shade calculates the skewness and similarity to GLCM. The larger the cluster shade intimates greater asymmetry with reference to average.

(v) The cluster tendency calculates a lot of centroids with the same gray level values. The contrast calculates the local magnitude variation, the values far from the cross diagonal. The bigger values correlate with higher divergence in the magnitude values between the near centroid.

(vi) The correlation values between 0 and 1 display the linear need for gray level values to their particular centroid on the gray level co-occurrence matrix.

(vii) The difference average calculates the connection between the occurrence of two identical pixel values and the occurrence of two varying intensity values.

(viii) The variance difference calculates heterogeneity which places larger weights on varying gray levels put together that diverge from the average.

(ix) The difference entropy calculates the variability in the nearest intensity value differences.

3.6.2 SPEEDED-UP ROBUST FEATURES

SURF is an algorithm that is quick and robust for local, similarity-invariant image comparison and representation (Allaberdiev et al., 2019). The concept of this method is based on the rapid calculation of work using box filters that help in enabling real-time processes like object tracking and object recognition, motion-based segmentation, and 3D reconstruction (Bay et al., 2006). These techniques worked in three ways: feature extraction, feature description, and feature matching. For locating locations of interest, the SURF employs a BLOB detector based on the Hessian matrix, applying sufficient Gaussian weights to wavelet responses in both horizontal and vertical directions for orientation assignment. SURF employs wavelet responses for feature description as well. A neighborhood surrounding the keypoint is picked and subdivided, and then the wavelet responses are extracted and represented for each subregion in order to generate the SURF feature descriptor (Harris and Stephens, 1988).

$$M = \sum_{x,y} w(x,y) \begin{bmatrix} I_x^2 & I_x I_y \\ I_x I_y & I_y^2 \end{bmatrix} \qquad (3.4)$$

where M is a second-order moment matrix obtained from image derivatives, $w(x,y)$ represents the window function, I_x indicates the image gradient in the horizontal direction, and I_y represents the image gradient in the vertical direction.

3.6.3 Gray Level Run Length Matrix

GLRLM is mainly used in medical imaging for statistic recording distribution and the relationship between image pixels. This feature extraction method was introduced by Galloway in 1975. With the same intensity values, the pixels can be described in a certain direction in a gray level run matrix. The pixels are called the gray level and the number of occurrences is called run-length values. Like GLCM, a number of parameters can be extracted which are as follows: short-run emphasis, long-run emphasis, and gray level non-uniformity. Some of the features are detailed as follows:

(i) Short-run emphasis which calculates the disbursement of small-run length, with a larger value representing smaller run lengths and smooth textural texture.

(ii) The long-run emphasis which calculates the disbursement of long-run lengths, with larger values representing the bigger run lengths and rough structural texture.

(iii) Gray level non-uniformity calculates the pixel values in the image, the smaller gray level based on non-uniformity values correlates with a larger intensity value.

(iv) The gray level non-uniformity normalized computes the likeness of the pixel value in the pictures, smaller the gray level non-uniformity normalized values correlate with a larger likeness in the intensity values.

(v) The run-length non-uniformity computes the likeness of the run lengths for the whole image; the lower values represent more uniformity between the run length in the pictures.

(vi) Run-length non-uniformity normalized computes values from the run-length image, the lower values represent more uniformity between the run lengths in the pictures.

(vii) The Run percentages calculate the roughness of the texture by using the number of ratios of runs and the number of centroids in the required ROI.

(viii) The runs in gray level intensity calculate the variance in the gray level variance.

(ix) The run from the run length calculates the variance in the run variance.

(x) The low gray level run calculates the disbursement of low gray level value with larger values representing a great focus on low gray level value in an image.

(xi) The high gray level emphasis calculates the disbursement of the larger gray level value with a higher value specifying a greater focus on higher gray level values in an image.

(xii) The short-run low gray level emphasis calculates the continuous distribution of lower run length along with the lower gray level values.

(xiii) The short-run high gray level emphasis which calculates the continuous distribution of short-run length with the high gray level values.

(xiv) Long-run low gray level run emphasis measures the continuous distribution of large run lengths with the lower gray values.

(xv) The long-run high gray level emphasis calculates the long-run lengths along with high gray level values.

3.6.4 Local Binary Pattern

The local binary pattern is a simple and well-organized texture operator that may be identified as pixels of an image by thresholding of eight adjacent pixels; the resulting numbers will be one and zero (Nanni et al., 2010). It is a robust feature for texture classification, provides local image properties, is coupled with histograms to increase detection performance in specific datasets, and can describe any image with a simple data vector. It uses four parameters: radius, neighbors, Grid X, and Grid Y. The parameter is used to find the binary values in which Grids X and Y provide the histogram of an image. The well-organized LBP employs the local spatial approach, and the grayscale image has contrast. The LBP texture descriptors use the segmented image's pixel value and distinguish the center pixel value to get the threshold values. Using the threshold values, features were retrieved.

3.6.5 Gray Level Size Zone Matrix

The GLSZM is derived from the GLRLM. The gray level size zone matrix deals with the number of size (s) and gray level (g) (Table 3.1). The matrix is calculated from the integer lines identical to n, the integer of gray level, the active number of columns, the size of the biggest level, and the amount of quantization. Compared to GLCM and GLRLM, GLSZM is a rotation-free representation of

TABLE 3.1
Gray Level Size Zone Matrix

Gray Level	Size Zone			
	1	2	3	4
1	1	2	0	0
2	2	0	2	0
3	0	0	1	0
4	1	0	1	0

connected gray level values and associated zones or areas in the ROI. Some of the features extracted from the GLSZM matrix are described as follows (Fusco et al., 2021):

(i) The small area emphasis describes the small size zones from the greater values for fine textures.
(ii) The large area emphasis describes the large zones from the greater values with large size zone and coarse textures.
(iii) The gray level non-uniformity takes a flexibility gray level intensity value from the picture. The smaller values give the uniformity values.
(iv) Size zone non-uniformity computes the flexibility of zone volume dimension in the image, with the bottom-most pixel value having more uniformity.
(v) Zone percentage calculates the roughness of the texture by integrating the ratio of integers of the zone along with the number of centroids in the ROI.
(vi) The gray level variance calculates the variance in gray level force for zones.
(vii) The zone variance computes variance in the number of volumes from the zone.
(viii) Low gray level zone emphasis calculates the disbursement of smaller gray level quantity zones, with a high value representing the great proportion of low gray level values and quantity of zones in the picture.
(ix) Zone entropy calculates the uncertainty in the disbursement of zone sizes and the gray level. The high values represent more uniformity in texture patterns.
(x) The high gray level zone emphasis calculates the disbursement of larger gray level value, for the larger values represent a greater measure of the high gray level value and the size zone in the picture.
(xi) The Small area low gray level emphasis calculates the measure of images in the continuous disbursement of small size zones for the low gray level values.
(xii) The small area high ray level emphasis measures in images of the continuous distribution of small size zones for high gray level values.
(xiii) The large area low gray level emphasis calculates the measure in an image of the continuous disbursement of large size zones for low gray level value.
(xiv) The large area high gray level emphasis calculates the proportion in an image of the continuous disbursement of the large number of zones for high gray level value.

3.6.6 Local Binary Gray Level Co-occurrence Matrix

This technique is based on a combination of LBP and GLCM algorithms (Öztürk and Akdemir, 2018). The features are extracted by taking all texture structures and spatial information. First, the local binary pattern creates a texture image, then the GLCM features were generated in the local binary pattern image. The GLCM takes only the pixel and neighbor pixel values; the advantage of the LBGLCM is extracting not only the pixel values but also the local pattern of the images by using the LBP algorithm. Nowadays, LBGLCM is widely used in image processing applications for a better understanding of the texture features. The gray scale texture calculates from the analysis of 3 × 3 local descriptor above the innermost pixel. It creates the keypoint by thresholding a local descriptor by the gray values of the central pixel, based on a local texture pattern. The eight surrounding pixels were assigned binary values, and these values were compared to the values of the core pixel. For example, gray values that are low or below the center pixel's gray value are designated as 0; otherwise, it will be 1.

3.6.7   SCALE-INVARIANT FEATURE TRANSFORM

SIFT algorithm was developed in 2004 by D. Lowe. The algorithm works in four ways of scale-space pixel selection: keypoint localization, orientation assignment, keypoint descriptor, and keypoint matching (Lowe, 2004). Scale-space of a picture is given by a task M (x, y, α) which is created from the convolution of Gaussian kernel in varying scales with the input picture, where (x, y) represents the pixel coordinates of Gaussian space and α indicates the smoothening parameter. The scale-space is split into octaves and the integer of octaves and scale depends also on the size of the actual image. A SIFT feature is a chosen picture area (also known as a keypoint) accompanied by a description. The SIFT detector extracts keypoints and the SIFT descriptor computes their descriptors. It is also typical to employ the SIFT detector and the description separately. A SIFT keypoint is a circular, oriented picture area. It is characterized by a geometric frame with four parameters: x- and y-coordinates of the keypoint's center, its scale (the region's radius), and its orientation. As keypoints, the SIFT detector employs picture structures that resemble "blobs." The SIFT detector is insensitive to translation, rotation, and rescaling of the picture since it searches for blobs at different scales and places.

3.6.8   LOCAL DIRECTIONAL PATTERN

LDP is a gray scale texture pattern that specifies the non-linear form of local images (Hasanal Kabir et al., 2012). LDP is a texture pattern that defines the spatial organization of a local picture texture in grayscale. An LDP operator computes the edge response values in each of the eight directions at each pixel point and creates a code based on the relative strength magnitude. Since edge responses are less susceptible to light and noise than intensity values, the resulting LDP feature characterizes local primitives such as various kinds of curves, corners, and junctions more accurately and maintains more information. LDP features are based on binary codes of eight bits that are allocated to each pixel of an input picture (Sargent et al., 2009).

3.6.9 SEGMENTATION-BASED FRACTAL TEXTURE ANALYSIS

The texture features and fractal analysis of image texture are extracted from an image using the SFTA algorithm (Öztürk and Akdemir, 2018). The SFTA contains two function variables I and nt: I is the input of gray scale picture and nt represents the amount of the feature vector. The gray level images are used as the input image, the multi-level threshold is applied, the input images are taken as the binary input image. The two methods are threshold binary decomposition (TTBD) and Fractal measurement. It is used to learn the boundary complexity of the object and the structure in the image. The segmented structure and edge complication of the object are described by the input image of fractal measurements. The edges of binary images are marked as x and y. The pair of eight connected pixels value is 1 at a position responding to the binary image had value one and one neighbor pixel values as zero, or else it will be taken as zero. The resulting edge is one pixel. D fractal dimension is calculated from all edge pictures by using the box-counting method. The total gray level and the amount of pixel count round off the detailed extraction of all binary images in the absence of extensively improving the calculation time. The TTBD takes gray scale images as input and then returns them as a pair of binary images.

3.7 MACHINE LEARNING CLASSIFIERS

Machine learning classifiers use the features extracted in the feature extraction process to understand how the features are related to the classes classified. The algorithms are capable of automatically classifying data into two or more classes, depending upon the attributes or features that are used as the input to the classifiers. Some of the most used machine learning classifier algorithms in literature are as follows:

- Logistic regression
- Naïve Bayes
- K-Nearest Neighbors
- Decision Tree
- Support Vector Machines
- Bagging
- LogitBoost
- K-star

3.7.1 LOGISTIC REGRESSION

Logistic calculation is used to predict a binary output. It evaluates discrete values (binary values like 0 and 1, yes or no, true or false, etc.) based on the independent variables. The output independent variables are analyzed to be a binary result which falls into two categories. The independent variables can be categorical or numeric, and at the same time, the dependent variables will be categorical. It is also used to calculate the positive and negative between zero and one and is also used to predict the number of different objects in a picture.

$$P(a = 1|b) \text{ or } P(a = 0|b) \tag{3.5}$$

P represents the conditional probability which indicates the occurrence of 0 or 1 based on the independent variable, where "*a*" indicates the dependent variable and "*b*" indicates the independent variable.

3.7.2 Naïve Bayes Classifier

This algorithm is based on the Bayes theorem which is based on the concept of conditional probability. Conditional probability is the estimation of the probability of one event, given that another one has already occurred. The Naïve Bayes is the fastest algorithm when compared to other algorithm techniques and is also relatively simple and easy to implement. This algorithm needs a small amount of training data to evaluate the necessary parameters. The Bayes algorithm is a supervised learning algorithm. It is possible to apply it in both binary and multiclass classifications. The Bayes algorithm can be mathematically represented as follows:

$$P(a|y) = p(y|a) p(a) / P(y) \tag{3.6}$$

where $p(y|a)$ = Likelihood, $p(a)$ = Class prior probability, $P(a|y)$ = Posterior probability, and $P(y)$ = Predictor probability.

3.7.3 K-Nearest Neighbors

k-nearest neighbors are commonly known as *k*-NN and are one of the simplest algorithms which is easy to apply. It is a supervised machine learning algorithm that can be used to solve both regressions as well as classification-related problems (Erickson et al., 2017). The *k*-NN algorithm uses data points to classify new data points based on the similarity of the points. *k*-NN algorithm is a data categorization approach that estimates the probability that a data point will become a member of one group or another depending on which group the data points that are closest to it belong to. The *k*-NN method is a supervised machine learning technique that is used to address classification and regression issues. However, its primary use is in categorization difficulties. *k*-NN uses a voting process to decide the category of an unseen observation. This indicates that the class with the most number of votes will be assigned to the data point in question.

3.7.4 Decision Tree

Decision tree algorithms are built in the same way as a big tree which contains a root (stem), branches, and leaves. Given data of attributes are joined along with classes; a decision tree creates a pair of rules that can be used to categorize the data. This algorithm is simple to interpret, requires less data preparation, and could be able to examine both categorical and numerical data. The decision tree nodes display the attribute, link branch, rule, and outcome. The structure of the decision tree leads to a data distribution that can be easily and unambiguously interpreted. The attribute set is taken as the input, then, it classifies the model, and finally labels the class as output.

3.7.5 SUPPORT VECTOR MACHINE

Support vector machines (SVMs) make use of hyperplanes in an N-dimensional space to classify the data points (Wernick et al., 2010). The number of hyperplanes corresponds to the number of classified features. The kernel trick technique transforms the data and with that transformation, it is possible to find the optimum edges between the outputs. For finding the margin region, the largest distance between the data points of both the classes will be taken. The data points are classified by decision boundaries called hyperplanes. The dimension depends upon the number of features. If the number of input features is one or two, it is easy to show that hyperplane is just a line. When the number of features increases, it is extremely hard to find the line. The support vector is a data point that is nearer to the hyperplane than the marginal plane of cluster class and takes the advantage of the place and orientation of the hyperplane. Here, the data points refer to a set of data localized in a defined site that either belongs to class A or class B. By this support vector, the classifier border line is maximized. The position of the hyperplane can be changed by removing the support vectors.

3.7.6 RANDOM FOREST

Random forest classifiers have different decision trees for various subsamples of the datasets. It can be used to solve complex machine learning problems without compromising the accuracy of the result. It can do both classification and regression. It is built up of many folds of decision trees, and the mean of all decision tree results is taken as the output in regression. For classification problems, the output is determined by a large number of votes. The random forest regression is simple regression in that values of dependent and independent variables are moved in the random forest model. A specific prediction is produced from separate trees in the random forest regression technique. The total prediction of the separate trees is taken as the output of the regression. Regression random forest is just opposite to the random forest classification in that the output is taken by mode of decision tree class. The linear regression and random forest regression are under the same concept, but the working function is different. In linear regression $X = Ay + B$, where, X is a dependent variable, y is an independent variable, A is the parameter, and B is a constant. For regression, it looks like just a black box.

3.7.7 BAGGING

Bagging comes under the Meta algorithm (an algorithm constructing the new algorithm often by combining or weighing the outputs from the set of other algorithms) which is used for statistical classification and regression. It improves the stability and accuracy of the learning. The smaller value of variance avoids overfitting. It is a model averaging technique which can apply base classifiers on random subsets of the real data and the collection of individual predictions to form a final prediction. The bagging resamples the real training dataset with the replacement of some other data that may be present multiple times while others are left unreplaced. The concepts of bagging are quite easy if we fix several independent models and their total predictions to make a model with a lower variance. It is not possible to fix the fully loaded independent models as it would require a large amount of data. With the approximate properties of bootstrap samples, it is possible to fix models that are nearly independent. Bootstrap sampling is a type of resampling that involves repeatedly drawing sample data from the original data source with replacements. Bagging, also called bootstrap aggregation, is designed to increase the stability and accuracy of machine learning algorithms used in classification and regression. It is specifically used to reduce the variance within the noisy data set. The new bootstrap sample will function as another independent dataset from the true distribution by making the multiple bootstrap samples. Now, we can fix a weak learner from the sample by calculating the mean from the output. The mean of the weak learner's outputs never changes the output answer, but it can minimize the variance.

3.7.8 LOGIT BOOST

Formulated by Jerome Friedman, Trevor Hastie, and Robert Tibshirani, in the LogitBoost algorithm, as the name suggests, the aim is to attain the highest accuracy by boosting the classifier. LogitBoost is a classification boosting technique. LogitBoost and AdaBoost are comparable in that they both do an additive logistic regression. LogitBoost reduces the logistic loss whereas AdaBoost minimizes the exponential loss. It constructs a logistic model by iterative refinement, gradually adding more variables as new linear models are introduced. The concept is to automatically build a tree structure by recursively dividing the iterative fitting process into decision tree branches matching subsets of the data (Friedman et al., 2000).

3.7.9 K-STAR

K-star classifier is an instance-based classifier that can classify an instance by comparing it with already classed examples. The fundamental assumption of this classifier is that identical instances would have the same classifications. In instance-based classifiers, the distance function plays a major role in determining the similarity between the two instances. In the K-star classifier, distance is measured as a function of the entropies of the instances. Therefore, we can say that the K-star classifier differs from other instance-based classifiers in the distance function aspect as it uses entropy as a function of distance. Mathematically, the K-star prediction model can be represented as

$$P^*(C|x) = \sum_{y \in c} P^*(x|y) \qquad (3.7)$$

where "x" and "y" are the two instances, and we must predict the probability of instance "x" belonging to category "C." This is done by summing up the probabilities of instance "x" with every other instance that belongs to "C." Some common hyper-parameters of the K-star classifier include "weights," which sets the weights for each distance, and "distance" and "metric" parameters, which decide how the distance is to be calculated from the neighboring point to an unknown point.

3.8 DEEP LEARNING CLASSIFIER

Deep learning has proven to be a particularly useful technique in recent decades due to its capacity to manage massive volumes of data. Increasing the number of neural processing layers (deep neural network (DNN)) has shown state-of-the-art performance compared to traditional techniques in popularity as well as in pattern recognition tasks. CNNs are one of the most widely used DNNs for computer vision problems. The earlier work could be dated back to the early 1960s when researchers struggled to create an intelligent system that can grasp visual input. This field became known as "computer vision" in the years that followed (Russakovsky et al., 2015). CNN-based computer vision reached a global status when a team of researchers from the University of Toronto built an AI system that outperformed the top image recognition algorithms of that time by a considerable margin in 2012. CNNs have ever since emerged as a completely essential part of many computer vision applications. So, let's discuss CNN in detail.

3.8.1 BACKGROUND

CNNs are the forerunners in the neural network spectrum. They are networks which are designed to work with images. The prime concept behind CNNs is inspired by the working of the human retina. In the retina, every neural group is accountable for a single patch of the complete image.

While working on Sensory processing Dr. Hubel and Dr. Wiesel (Hubel and Wiesel, 1968) inserted a micro-electrode into a visual cortex of a cat and displayed line images at various angles to the cat. They found out that different neural bundles responded differently to the angles of the image. This work is the prime motivation behind the development of CNNs.

Image processing problems have a few significant properties that influence how they should be solved. The first is that images are massive in terms of input – in a fully connected network, a 200 × 200 image would feed 40000 individual weights to each node. The other is that location is important. For an extracted feature, pixels near the feature's central location in an image are selected and ones which are away are not selected. Because of this, a fully connected network is not an appropriate choice because it will be computationally uncontrollable.

CNN solves this issue by reducing the number of weights that each node must learn in an intelligent and locally aware way. An important basic feature of CNNs is that they do not vary from location to location.

3.8.2 BASIC ARCHITECTURE

A typical CNN architecture has the following attributes ("Deep Learning|The MIT Press" 2016):

- Convolution filters or receptive fields
- Pooling or Subsampling
- Fully connected layer (FCL)

Understanding how these processing layers work is a significant step in developing a good understanding of CNN because they are the basic building blocks of every CNN power complex computer vision application. We'll try to understand the reasoning behind each of these operations in the sections that follow.

3.8.2.1 Convolution Layer

In the case of a ConvNet or receptive field or convolution filter (used interchangeably), the main objective (Ghosh et al., 2020) of convolution is to extract textural features from the input image. By learning image features from smaller sections of input data, convolution preserves the spatial relations between pixels (Pattanayak, 2017).

As previously stated, each image can be understood as a matrix of pixel values. Consider a sample image with a 10 × 10 matrix with only 0- and 1-pixel values.

Also consider another 5 × 5 matrix which is the convolution filter. We slide this 5 × 5 matrix by one pixel over our original image and compute pixel-wise multiplication and add the multiplication outputs to get the actual integer, which creates a single element of the output sequence.

Mathematically, the convolution operation on the image can be represented as

$$f_g^h(k,l) = \sum_C . \sum_{y,z} i_C(y,z) * e_g^h(s,t) \tag{3.8}$$

where "k" and "s" represent rows whereas "l" and "t" depict columns under consideration. "$f_g^h(k,l)$" represents k, l element of feature matrix. "y" and "z" represent the x- and y-coordinates of the image. Channel index, represented as "c"

and $i_C(y, z)$, is a component of the input image tensor "i_C." "$e_g^h(s,t)$" represents the index of the hth convolution kernel h_g of the gth layer. The output of the hth convolution function can be written as

$$F_g^h = \left[f_g^h(1,1), \ldots f_g^h(k,l), \ldots f_g^h(K,L) \right] \tag{3.9}$$

where "F_g^h" is the output of the convolution which is the input feature matrix for gth layer and hth neuron.

3.8.2.2 Rectified Linear Unit (ReLU)

This is typically applied element-by-element to the output of another function, such as a matrix-vector product. The activation function for F_g^h can be described as

$$T_g^h = \eta_a\left(F_g^h\right) \tag{3.10}$$

where η_a depicts an activation function and T_g^h is a transformed output.

Rectified Linear Unit (ReLU) operation substitutes the negative pixel values in the feature map with zero. Hence, the goal of ReLU is to include non-linearity into the CNN model, as most of the actual data we want our CNN to learn is non-linear in nature.

3.8.2.3 Pooling Layer

Pooling (Guo et al., 2015) also known as subsampling or downsampling process reduces the dimensionality (activation map size) of each feature vector while retaining the most significant data. The pooling layer can be represented as

$$R_g^h = \eta_p\left(F_g^h\right) \tag{3.11}$$

where "η_p" represents the pooling operation and "R_g^h" is the feature map (pooled) of gth layer for hth input feature F_g^h. Various types of pooling formulations such as overlapping, max, spatial pyramid pooling, average, L2, etc., can be used in CNN to extract a combination of features.

3.8.2.4 Max Pooling

Max pooling is a subsampling process that selects the maximum value from the activation or feature map. As a consequence, the output of the max-pooling layer would be a feature map containing the most significant features from the earlier feature map.

3.8.2.5 Average Pooling

Average pooling is also a type of pooling operation whereby it calculates the average of the elements in the filter's activation map region. As a result, when comparing the max-pooling and average-pooling operations, the former returns the most distinctive feature in a given patch of the feature map; however, the latter returns the average of the features present in that patch.

3.8.3 Batch Normalization

Batch normalization acts as a normalizing factor as it eases the movement of the gradient, thus helping in advancing the generalization of the network represented as

$$N_g^h = \frac{F_g^h - \mu_A}{\sqrt{\sigma_A^2 + \varepsilon}} \qquad (3.12)$$

where N_g^h and F_g^h represents the normalized feature map and input feature map, respectively. Whereas μ_A and σ_A represents the mean and variance, respectively, and ε is included for mathematical stability to avoid division by zero.

3.8.4 Drop Out

Drop out initiates regularization which improves generalization within the network. It works by randomly omitting some connections through a reliable probability which produces numerous thinned network architectures. Thus, selecting one representative network with small weights will be then considered as an estimation of the entire proposed network.

3.8.5 Fully Connected Layer

FCL is a feedforward neural network and is present at the end of the network for classification. Here, the activation maps generated from feature extraction stages act as input, and evaluation of the output obtained from all the previous CNN layers takes place. Finally, a non-linear grouping of selected features is made, which is intended for the classification of data.

3.8.6 Training Procedure for CNN

The training procedure consists primarily of the following steps ("Deep Learning | The MIT Press" 2016):

- Pre-processing and data augmentation
- Initialization of parameters
- Regularization
- Selection of an optimizer

3.8.7 Pre-Processing and Augmentation of Data

Data pre-processing (Patil and Rane, 2021) is the application of transformations to raw data in order to make the dataset more amicable, feature-rich, learnable, and uniform for CNN layers. Before feeding the data to the CNN model, the data is pre-processed.

Data augmentation is used for increasing or expanding the size of the training dataset artificially. In this step, we apply various image- or data transformations to data samples from the training dataset only and convert them into new data samples, which are then used for training the CNN models.

Data augmentation is also significant because, for most real-world problems, a very inadequate-sized training data set is available, and the truth is that the more the training data samples available, the better the CNN model. Some of the image transformations that are widely used in literature when performing image classification tasks are cropping, translations, flipping, contrast adjustment, scaling, rotations, and other advanced data augmentation operations based on generative models are also available.

3.8.8 INITIALIZATION OF PARAMETERS

A typical CNN architecture is made up of millions of parameters. As a result, it must be properly initialized at the onset of the training process, because the initialization of weight directly dictates how accurately and quickly the CNN model will converge to the optimal setting (Pattanayak, 2017). The most straightforward method is to set all the weights to zero. However, this turns out to be a major error because if we set the weights of all layers to zero, the output, as well as the gradients calculated by each neuron in the network during backpropagation, will be the same. To avoid this discrepancy among neurons, we do not set all weights to the same value at the start. This is achieved by two methods, namely random initialization and unsupervised pre-training.

3.8.8.1 Random Initialization

As the name implies, we use random matrices to randomly initialize the weights, with the elements of those matrices sampled from some probability distribution with a small standard deviation and zero mean. The main issue with this method is that it can result in vanishing gradients.

3.8.8.2 Unsupervised Pre-Training Initialization

In this technique, we initialize the weights of the CNN with another CNN which is trained with an unsupervised technique, such as a deep belief network.

3.8.9 CNN REGULARIZATION

The ability of deep learning algorithms to evolve adequately to new and relatively unseen input taken from the same distribution as training data is known as generalization. Over-fitting is the main issue that prevents a CNN model from achieving good generalization. This type of error occurs when a model performs extraordinarily well on training data but fails on test data. The inverse is an under-fitted model, which occurs when the model has not started learning enough from the training data. When the model performs excellently both for train and test data, it is referred to as a just-fitted model.

3.8.10 CHOOSING AN OPTIMIZER

Optimizers are algorithms or methods that are used to determine the characteristics of your neural network, such as weights and learning rate, in order to improve efficiency. The learning rate is the tuned hyperparameter that controls how much the model has to change in response to estimated error when each time the model weights are updated. The target parameter of efficiency is defined as more precise defined weights in a short period of time.

3.8.10.1 Gradient Descent

Gradient descent is the simplest but most widely used optimization algorithm. It is extensively used in linear regression and classification algorithms. A gradient descent algorithm is also used in neural network backpropagation. Gradient descent is an optimization algorithm that is based on the first-order derivative of a loss function. It determines how the weights should be changed so that the function can reach a global minimum. Backpropagation transfers the loss from one layer to the next, and the model's parameters, also known as weights, are modified based on the losses so that the losses are minimized.

3.8.10.2 AdaDelta

It is an AdaGrad extension that attempts to eliminate the decaying learning rate problem. Rather than accruing all previously squared gradients, Adadelta restricts the window of accumulated past gradients to a fixed size given by "w." Instead of the sum of all gradients, an exponentially moving average is used in this case.

3.8.10.3 Adam

Adam (adaptive moment estimation) works with first- and second-order momentums. The notion behind Adam is that we do not want to update the weights too fast just because we can leap over the minimum; instead, we want to slow down the updating of weight for a more cautious search so that the global minimum is achieved. Adam, like AdaDelta, keeps an exponentially decaying average of past squared gradients, but it also keeps an exponentially decaying average of past gradients.

3.9 PERFORMANCE METRICS

3.9.1 CONFUSION MATRIX

The confusion matrix is one of the frequently used and simple metrics for determining the model's correctness and accuracy. It is used for classification problems with two or more types of classes as output.

- **True Positives (TP):** True positives occur when the actual class of a data point is true and the predicted class is also true.
- **True Negatives (TN):** True negatives occur when the data point's actual class is false and the predicted class is also false.

- **False Positives (FP):** False positives occur when the actual class of a data point is false but the predicted class is true. False because the model predicted incorrectly, and positive because the class predicted correctly.
- **False Negatives (FN):** False negatives occur when the actual class of a data point is true, but the predicted class is false. False because the model predicted incorrectly, and negative because the predicted class was negative.

3.9.2 ACCURACY

In classification problems, accuracy is defined as the number of correct predictions made by the model over all types of predictions made.

$$\text{Accuracy} = \frac{TP + TN}{TP + FP + TN + FN} \tag{3.13}$$

When the target variable classes in the data are nearly balanced, accuracy is a good measure.

- **Precision**: The precision is determined as the ratio of the number of positive samples correctly classified to the total number of positive samples classified either correctly or incorrectly. The precision metric assesses the model's classification accuracy.

$$\text{Precision} = \frac{TP}{TP + FP} \tag{3.14}$$

- **Recall or Sensitivity**: The recall is computed as the ratio of positive samples that were classified correctly to the total number of positive samples. The recall metric assesses the model's ability to identify positive samples. The greater the recall, the greater the number of positive samples detected.

$$\text{Recall} = \frac{TP}{TP + FN} \tag{3.15}$$

- **Specificity**: Specificity is described as the proportion of true negatives that were predicted as negatives. This suggests that there will be a proportion of actual negatives that were predicted to be positive, which could be referred to as false positives.

$$\text{Specificity} = \frac{TN}{TN + FP} \tag{3.16}$$

- **F1 Score**: The F score, also known as the F1 score, is a model's accuracy on a dataset. It is used to assess binary classifications that categorize instances as "positive" or "negative." The F score is a method of combining the model's precision and recall.

$$F1\,\text{Score} = \frac{TP}{TP + 1/2\left(FP + FN\right)} \tag{3.17}$$

CNN architectures are the most sought methods for image classification problems. Many recent research articles utilized CNN for image classification and the results obtained are also astonishing. Training a CNN model from the start is a very tedious task and requires a lot of time and a large data set. Therefore, the transfer learning approach is used wherein pre-trained models are utilized for image classification.

Although similar in concept and calculation, performance measures of classifications must not be confused with the accuracy of medical diagnostic tests. High accuracy in the classification of features does not confer information that the obtained accuracy coincides with the performance for detecting disease.

3.10   PRE-TRAINED NETWORK MODELS

Application of CNN architecture in computer vision problems such as image classification and object detection has been reported with state-of-the-art (SoA) performance because of the availability of large image databases such as ImageNet for training and validation of the deep learning models. ImageNet challenge contest which is held yearly since 2010 is termed as ImageNet Large Scale Visual Recognition Challenge (ILSVRC) for validating object and image classification models. The database provides millions of images which are grouped into 1000 image classes for performance analysis. One of the earlier CNN models which received SoA performance was the AlexNet CNN model which used an eight-layer (five convolution layers and three max-pooling layers) architecture for image classification (Krizhevsky, Sutskever, and Hinton, 2017). Later, many CNN models won the challenge which is based on top-5 or top-1 classification error rate (%) (Russakovsky et al., 2015). The CNN model architecture and weights were made public for the researchers to apply for their custom image classification tasks, hence the term pre-trained models. The models could be used for image classification tasks as such or tuned for required performance in computer vision problems arising in various fields such as medical imaging, astronomy, human monitoring, and surveillance (Rajinikanth et al., 2020). Section 3.10.1 explains some of the widely used CNN pre-trained models for common image classification tasks.

3.10.1   VGG Architecture

Visual Geometry Group (VGG) is a Very Deep Convolutional Network for Large-Scale Image Recognition, which was one of the earlier works that studied the impact of varying the depth of the convolution layers for the image classification problem. VGG won the ImageNet challenge 2014 with SoA performance compared to previously published CNN architectures. They were able to achieve superior classification accuracy by increasing the CNN layers (typical VGG architecture has 16 and 19 layers) and also reducing the computation time by going for smaller receptive fields, namely 3×3 convolution layers with a stride of 1. This was followed by max-pooling layers for some of the convolution layers. Finally, the network consists of feed-forward neural networks for high-level decision-making, and classification is done in the last layer with a softmax activation function. The trained VGG model could be used for general image classification and could also be modified based on transfer learning for custom image classification tasks (Simonyan and Zisserman, 2014).

3.10.2   RESNET V2

ResNet V2 is an ANN that is widely used for image classification and regression. It utilizes the skip connection mechanism by jumping over some layers. The main reason for skipping is to avoid the vanishing gradient problem by making use of activations from a previous layer until the next layer learns its weight. This residual learning and skip connections enable the ResNet V2. In ResNet, each convolution is followed by batch normalization and non-linear activation functions. The ResNet V2 net empowers the features which result in increased accuracies during the classification. The pre-trained ResNet V2 is modified using the transfer learning approach where the earlier layers are fixed to overcome the overfitting issues and the higher-level portions of the pre-trained net are fine-tuned for this current study. In the modified ResNet V2 net, the filter convolution layer and the inception ResNet layers are fixed and the later layers are fine-tuned by adding a global averaging pooling (GAP) layer followed by three FCLs and a softmax layer.

The input image of size $128 \times 128 \times 3$ is passed through the convolution layer where bytes normalization is followed by the application of the ReLU activation function, and then a convolution operation is executed on the feature maps. The ResNet V2 net contains three different types of inception modules, namely Inception ResNetA, Inception ResNetB, and Inception ResNetC. The inception modules help to generate discriminatory features and minimize the number of input parameters where each inception module consists of numerous convolutional and pooling layers in parallel. The feature maps obtained from the last inception layer are then passed through the GAP layer and an output of 2048 feature maps is obtained. Finally, three FCLs followed by the softmax layer are used to provide the classification (He et al., 2015).

3.10.3   DENSENET 121 (DENSELY CONNECTED CONVOLUTIONAL NETWORKS)

DenseNet 121 is a widely used net as it requires only minimal parameters to train the model. DenseNet improves gradient propagation by connecting each layer directly with every other layer in a feed-forwarded manner (Gao et al., 2017). The DenseNet architecture uses a residual mechanism to its full potential by making each layer connect to its subsequent layers. The thermogram's input ($128 \times 128 \times 3$) is given to the convolution layer where the input data gets multiplied by a filter and two-dimensional outputs (feature maps) are generated. The feature maps then pass through a ReLU activation layer followed by a convolution operation. The feature maps generated through the convolution layer pass through the dense block while executing convolution operation with a definite number of filters (k), where k is the growth factor. The output (feature maps) is generated by the dense block after concatenating to the initial feature map obtained from the convolution block. The generated feature maps are then passed through the transition layers that are used for downsampling the feature maps by applying batch normalization, a 1×1 convolution followed by 2×2 pooling layers. The feature maps generated from the last dense block are given as an input to the GAP layer. GAP layer is used for reducing (halved) the spatial dimension of the incoming feature maps. The output from the GAP layer is given to the three fully connected layers (FCL), namely FCL1, FCL2, and FCL3 having 128, 64, and 32 feature maps, respectively. Finally, the feature maps generated from the last FCL are passed through the softmax layer which provides the classification (Huang et al., 2018).

3.10.4 MOBILENET

Many pre-trained models discussed here, such as ResNet and DenseNet, are DNNs with many hidden layers. As the number of convolution layers increases, the computation cost of determining the weights of the network also increases. Hence, dedicated processing units are required to compute the parameters of the model. So, the application of these CNN models could not be used for device intelligence for embedded or mobile systems. MobileNet overcomes this problem, by introducing depth-wise separable CNN layers that lead to the estimation of parameters in real-time even for mobile devices. This task is achieved by factorizing the CNN layers into two stages of computation layers, namely (i) depthwise convolution and (ii) 1×1 pointwise convolution layers. The factorization of the CNN layers provides a huge reduction in the computation cost of the CNN model (Howard et al., 2017).

3.11   CONCLUSION

This chapter has provided in-depth knowledge about various steps involved in the image processing and classification of thermal and visible images for various image processing applications. Here, we have illustrated the methods of processing multispectral images such as thermal and visible images of a scene and also explained the different types of features that can be used for image classification. Moreover, this chapter also provides the recent trends in DNN and pre-trained CNN architectures that are extensively used not only in thermal image classification but also in image or object recognition.

REFERENCES

Allaberdiev S, Yakhyoev S, Fatkhullayev R, Chen J Speeded-up robust feature matching algorithm based on image improvement technology. *Journal of Computer and Communications* 2019, 7, 1–10. https://doi.org/10.4236/jcc.2019.712001

Batchuluun G, Yoon HS, NGuyen DT, Pham TD, Park KR. A study on the elimination of thermal reflections. *IEEE Access* 2019, 7, 174597–174611. https://doi.org/10.1109/access.2019.2957532

Bay H, Tuytelaars T, and Van Gool L. 2006. "SURF: Speeded Up Robust Features." In *Computer Vision – ECCV 2006*, 404–17. Berlin, Heidelberg: Springer.

Erickson BJ, Korfiatis P, Akkus Z, Kline TL. Machine learning for medical imaging. *Radiographics* 2017, 37(2): 505–515.

Erwin E. "Peer Review Performance Analysis of Comparison between Region Growing, Adaptive Threshold and Watershed Methods for Image Segmentation." (2020).

Ester M, Kriegel H-P, Sander J, Xu X A density-based algorithm for discovering clusters in large spatial databases with noise. In E. Simoudis, J. Han, & U.M. Fayyad (Eds.), *Proceedings of the second International Conference on Knowledge Discovery and Data Mining*. Portland: AAAI Press. 1996.

Friedman J, Hastie T, Tibshirani R Additive logistic regression: A statistical view of boosting. *The Annals of Statistics* 28(2): (2000), 337–374.

Fusco R, Piccirillo A, Sansone M, Granata V, Rubulotta MR, Petrosino T, Barretta ML, Vallone P, Di Giacomo R, Esposito E, Di Bonito M, Petrillo A. Radiomics and artificial intelligence analysis with textural metrics extracted by contrast-enhanced mammography in the breast lesions classification. *Diagnostics* 2021, 11(5): 815. https://doi.org/10.3390/diagnostics11050815

Gao H, Liu Z, Van Der Maaten L, Weinberger KQ. Densely Connected Convolutional Networks. In: *Proceedings - 30th IEEE Conference on Computer Vision and Pattern Recognition, CVPR* 2017. https://doi.org/10.1109/CVPR.2017.243

Ghosh A, Sufian A, Sultana F, Chakrabarti A, De D Fundamental concepts of convolutional neural network. In: *Recent Trends and Advances in Artificial Intelligence and Internet of Things*. Springer, Cham, 2020. 519–567. https://doi.org/10.1007/978-3-030-32644-9_36

Gonzalez RC, Woods RE *Digital Image Processing*, 4th Edition, Pearson. 2018, Accessed August 27, 2021. https://www.pearson.com/us/higher-education/program/Gonzalez-Digital-Image-Processing-4th-Edition/PGM241219.html

Goodfellow I, Bengio Y, Courville A *Deep Learning*. The MIT Press 2016. Accessed August 27, 2021. https://mitpress.mit.edu/books/deep-learning

Guo G, Wang H, Yan Y, Zheng J, Li B A fast face detection method via convolutional neural network. *Neurocomputing* 2015, 395(Bo Li): 128–137. https://doi.org/10.1016/j.neucom.2018.02.110

Guo Y, Sengur A NCM: Neutrosphic c-means clustering algorithm. *Pattern Recognition* 2015, 48: 2710–2724.

Harris C, Stephens M A combined corner and edge detector, in *Alvey Vision Conference*, vol. 15, Manchester, UK, 1988, pp. 10–5244.

He K, Zhang X, Ren S, Sun J Deep Residual Learning for Image Recognition. *Proceedings of the IEEE Computer Society Conference on Computer Vision and Pattern Recognition 2016-December*, 2015 (December): 770–778. https://arxiv.org/abs/1512.03385v1

Hemalatha RJ, Thamizhvani TR, Josephin Arockia Dhivya A, Joseph JE, Babu B, Chandrasekran R Active contour based segmentation techniques for Medical image analysis. In: Robert Koprowski (Eds), *Medical and Biological Image Analysis*, 2018, https://doi.org/10.5772/intechopen.74576

Howard AG, Zhu M, Bo C, Kalenichenko D, Wang W, Weyand T, Andreetto M, Adam H *MobileNets: Efficient Convolutional Neural Networks for Mobile Vision Applications.* April. 2017. https://arxiv.org/abs/1704.04861v1

Huang G, Liu S, Van der Maaten L, Weinberger KQ Condensenet: An efficient densenet using learned group convolutions. In *Proceedings of the IEEE conference on computer vision and pattern recognition*, pp. 2752–2761. 2018.

Hubel DH, Wiesel TN Receptive fields and functional architecture of monkey striate cortex. *The Journal of Physiology* 1968, 195 (1): 215–243. https://doi.org/10.1113/jphysiol.1968.sp008455

John V, Tsuchizawa S, Liu Z, Mita S Fusion of thermal and visible cameras for the application of pedestrian detection. *Signal, Image and Video Processing* 2017, 11(3): 517–524. https://doi.org/10.1007/s11760-016-0989-z

Kabir H, Jabid T, Chae O. Local directional pattern variance (LDPv): A robust feature descriptor for facial expression recognition: The international arab *Journal of Information Technology* 2012, 9(4): 382–391.

Krizhevsky A, Sutskever I, Hinton GH ImageNet classification with deep convolutional neural networks. *Communications of the ACM* 2017, 60(6): 84–90.

Lowe DG Distinctive image features from scale-invariant keypoints. *International Journal of Computer Vision* 60: 91–110 (2004). https://doi.org/10.1023/B:VISI.0000029664.99615.94

Ma J, Liang P, Yu W, Chen C, Guo X, Wu J, Junjun JJ Infrared and visible image fusion via detail preserving adversarial learning. *Information Fusion* 2020, 54: 85–98. https://doi.org/10.1016/j.inffus.2019.07.005

Ma J, Ma Y, Li C. Infrared and visible image fusion methods and applications: A survey. *Information Fusion* 2019, 45: 153–78. https://doi.org/10.1016/j.inffus.2018.02.004

Mall PK, Singh PK, Yadav D GLCM based feature extraction and medical X-RAY image classification using machine learning techniques, *IEEE Conference on Information and Communication Technology*, 2019, 2019, pp. 1–6, doi: 10.1109/CICT48419.2019.9066263

Mittal H, Pandey AC, Saraswat M et al. A comprehensive survey of image segmentation: Clustering methods, performance parameters, and benchmark datasets. *Multimed Tools Application* 2021. https://doi.org/10.1007/s11042-021-10594-9

Muthukrishnan R, Radha M. Edge detection techniques for image segmentation. *International Journal of Computer Science and Information Technology*, 2011, 3. https://doi.org/10.5121/ijcsit.2011.3620 259

Nanni L, Lumini A, Brahnam S Local binary patterns variants as texture descriptors for medical image analysis, *Artificial Intelligence in Medicine*, 2010, 49: 117–125.

Öztürk S, Akdemir B Application of feature extraction and classification methods for histopathological image using GLCM, LBP, LBGLCM, GLRLM and SFTA. *Procedia Computer Science* 2018, 132: 40–46.

Pajares G, de la Cruz JM A wavelet-based image fusion tutorial. *Pattern Recognition* 2004, 37(9): 1855–72. https://doi.org/10.1016/J.PATCOG.2004.03.010

Panetta K, Wan Q, Agaian S, Rajeev S, Kamath S, Rajendran R, Wan Q, Agaian S, Rao SP, Kaszowska A, Taylor HA, Samani A, Yuan X, et al. A comprehensive database for benchmarking imaging systems. *IEEE Transactions on Pattern Analysis and Machine Intelligence* 2020, 42(3): 509–20. https://doi.org/10.1109/TPAMI.2018.2884458

Patil A, Rane M Convolutional neural networks: An overview and its applications in pattern recognition. *Smart Innovation, Systems and Technologies* 2021, 195: 21–30. https://doi.org/10.1007/978-981-15-7078-0_3

Pattanayak S *Pro Deep Learning with TensorFlow*. Packt Publishing Ltd, Apress, Berkeley, CA, 2017. https://doi.org/10.1007/978-1-4842-3096-1

Rajinikanth V, Raj ANJ, Thanaraj KP, Naik GR. A customized VGG19 network with concatenation of deep and handcrafted features for brain tumor detection. *Applied Sciences* 2020, (10): 3429. https://doi.org/10.3390/app10103429

Ronneberger O, Fischer P, Brox T U-Net: Convolutional networks for biomedical image segmentation, 2015, https://arxiv.org/pdf/1505.04597.pdf

Russakovsky O, Deng J, Su H, Krause J, Satheesh S, Ma S, Huang Z, et al. ImageNet large scale visual recognition challenge. *International Journal of Computer Vision* 2015, 115(3): 211–52. https://doi.org/10.1007/s11263-015-0816-y

Sander J, Ester M, Kriegel H-P, Xu X Density-based clustering in spatial databases: The algorithm GDBSCAN and its applications. *Data Mining and Knowledge Discovery* 1998, 2(2): 169–194.

Saparudin E, Nevriyanto A, Purnamasari D Performance analysis of comparison between region growing, adaptive threshold and watershed methods for image segmentation. *Proceedings of the International MultiConference of Engineers and Computer Scientists 2018*, 2018, Hong Kong.

Sappa AD, Carvajal JA, Aguilera CA, Oliveira M, Romero D, Vintimilla B. Wavelet-based visible and infrared image fusion: A comparative study. *Sensors* 2016, 16: 861. https://doi.org/10.3390/S16060861

Sargent D, Chen C-I, Tsai C-M, Wang Y-F, Koppel D. Feature detector and descriptor for medical images. *Proceedings SPIE 7259 Medical Imaging 2009: Image Processing*, 2009. https://doi.org/10.1117/12.811210

Sharma N, Aggarwal LM Automated medical image segmentation techniques. *Journal of Medicine and Physics* 2010, 35: 3–14. https://doi.org/10.4103/0971-6203.58777

Simonyan K, Zisserman A. Very deep convolutional networks for large-scale image recognition 2014 arXiv preprint arXiv:1409.1556,2014

Teledyne FLIR. n.d. "FLIR E54 Advanced Thermal Imaging Camera | Teledyne FLIR." Accessed August 27, 2021. https://www.flir.in/products/e54/

Vollmer M, Möllmann KP. *Infrared Thermal Imaging: Fundamentals, Research and Applications*, 2nd Edition, n.d., John Wiley & Sons, Germany.

Wernick MN, Yang Y, Brankov JG, Yourganov G, Strother SC Machine learning in Medical imaging. *IEEE Signal Processing Magazine* 2010, 27(4): 25–38.

Wu H, Huang J, Wang X, Jia J Image Registration of Infrared and Visible Based on SIFT and SURF. In: *Tenth International Conference on Digital Image Processing (ICDIP 2018)*, edited by Jiang X and Hwang J.N., 2018, 10806:186. SPIE. https://doi.org/10.1117/12.2503048

4 Thermal Imaging for Arthritis Evaluation in a Small Animal Model

Earlier researchers have proved that the animal model has served as a valuable tool to detect the progression of arthritis disease characteristics. The evaluations of pathogenesis and inflammatory arthritis in animal models and anti-arthritic drug development for arthritis treatment have been reported by Bendele. Among the various adjuvant-induced arthritis (AIA), the Complete Freund Adjuvant (CFA)-induced arthritis model resembles the features of arthritis and is used as the best model for the evaluation of chronic inflammation in arthritis.

Although various imaging modalities such as X-ray, MRI, ultrasound, and PET are available for diagnosis of rheumatoid arthritis (RA). Thermal imaging is considered a simple non-invasive screening tool for the assessment of arthritis in animal models and humans. The impactful feature of thermal imaging used in diagnostic mode is based on blood flow in a particular area which reflects the increase or decrease in temperature. The variation in temperature is directly correlated with some functional and pathological processes associated with the inflammatory reaction.

4.1 INDUCTION AND EVALUATION OF ARTHRITIS

4.1.1 COLLAGEN-INDUCED ARTHRITIS

Rats are immunized with the combination of type-II collagen in IFA or CFA for inducing arthritis. The presence of collagen-induced arthritis (CIA) in Th17 cells plays a predominant role in the pathological observation of arthritis. The histo-pathological results revealed the close resemblance of CIA with RA with regards to the cartilage and bone destruction and due to the presence of infiltrating cells in synovial tissue. The collagen type-II model is considered to be an important autoantigen in RA. This type of model induces inflammation in articular joints and causes joint destruction (Kim et al., 1999, 2004).

Jasemian et al. (2011) selected 16 adult female rats with weights ranging between 160 and 180 g. They injected porcine type II CIA in 16 rats. Then, they prepared an arthritis-inducing emulsion by adding equal quantity of collagen and Incomplete freund adjuvant (IFA) in a ice-water bath. Then the rats were immunized with 0.2 mL of the emulsion through subcutaneous injection at the base of the tail. After seven days of onset of arthritis inducing emulsion, the rats were then immunized with 0.1 mL of booster infection to elevate the severity of arthritis. Then, they divided the rats into two groups, namely the vehicle group (12 rats) and the control group (12 rats). The vehicle group was treated with sterile phosphate-buffered saline (PBS) whereas the control group was treated with dexamethasone. All the study rats were treated two times per week and the measurements were performed. Then, the surface temperature of the study rats was measured using a SATIR-S280 infrared video camera to monitor the temperature pattern. The skin surface temperatures at the hind feet (metatarsals and tarsus) were recorded. The scoring system followed in ankle and digits in the range varies from 0 to 4 are as follows: (i) 0 shows normal, (ii) 0.5 represents light redness, (iii) 1 indicates mild redness, (iv) 2 depicts moderate redness and swelling, (v) 3 shows severe redness and swelling of an entire paw, and (iv) 4 represents maximally inflamed limb in multiple joints. Then, they measured edema of the hind foot using a digital caliper and it was evaluated by measuring the cross-section area of the paw and ankle in each hind foot. The bodyweights of the rats were also measured along with the clinical scores. The measurements were made before treatment, i.e., on day 0 and on days 7, 9, 11, 14, 17, 21, and 23 after the onset of treatment. They also used ANOVA, unpaired t-test, and Mann–Whitney rank-sum test for the comparison between relative organ weights for both the study group.

Nosratiz et al. induced bovine type-II collagen in CFA by means of intradermal injection in mice. After 21 days, a booster injection with incomplete IFA was provided. Further to enhancing the arthritis, 2 mg/kg lipopolysaccharides from *Escherichia coli* mixed in 100 μL saline were induced. The disease severity of the CIA-induced model was assessed in the wrist and ankle joints of the forepaw and the hind paw, respectively, of female DBA1/J mice. The control group mice were injected with a sterile saline in the CIA group. The mean baseline values of parameters such as paw thickness, arthritis score, and thermal imaging parameters were obtained prior to the immunization process. The authors anesthetized the CIA mice with 5% isoflurane induction prior to the measurements of paw thickness and arthritis scoring. The arthritis severity is scored based on visual inspection on a scale ranging from 0 to 4 as follows: (i) 0 – normal, (ii) 1 – mild redness, (iii) 2 – moderate redness and swelling, (iv) 3 – severe redness and swelling in ankle foot and digits. The paw thickness was measured using the digital caliper in millimeters from day 1.

Sanchez et al. conducted their study in the CIA model using female Dark Agouti (DA) rats weighing 110–125 g. After acclimation, the rats were provided with proper food and water. They dissolved the bovine type II collagen in 2 mg/mL 0.01 N acetic acid and kept it for 12 h duration with temperature maintained at 39.2°F. Then, they mixed the dissolved collagen with IFA at 39.2°F in equal proportions and kept the solution in a 30 mL syringe. Each rat is injected with the collagen-ICFA emulsion of a total of 500 μg on day 0. 100 μL is injected into each region of the rat, hence the emulsion was injected in five different regions such as the base of both shoulders, two hips, and one on the neck region (backside). A booster injection similar to the first injection was administered to the rats on day 7. Among a total of 18 rats, 10 rats were immunized and 8 rats were kept as controls. Thermal imaging was acquired in the hind paw region from day 6 and the imaging process was continued throughout the disease progression at every 3–4 days intervals. Then, a water plethysmograph was used to measure the edema in the hind paw region. Thermal images were obtained from both right- and left-side hind paws due to the polyarthritic nature.

Chan et al. conducted the study in a CIA model and obtained approval from the Temple University Institutional Committee for their study. 50 µg of chicken type II collagen blended with IFA was injected into old male DBA/1 mice 6–8 weeks old. 0.5 mm needle was used for inducing arthritis in the murine model and injected with 2–3 mm depth for all the small animals. At the same time, 0.9% sodium chloride is injected into control rats. On day 21, the reaction was increased again with a 50 g injection of collagen in incomplete Freund's adjuvant. Arthritis developed in 70–80% of inoculated mice between days 40 and 55, depending on the individual mouse. Footpad edema and histopathology were used to determine the progression of the disease. To track the progression of inflammation, a digital caliper was used for the measurement of the thickness of footpad by two independent observers. The arthritis severity is scored based on the formula given as follows:

Thickness in the footpad

$$= \frac{\left(\text{thickness at the day of observation} - \text{thickness before immunization} \right)}{100}$$

$$(4.1)$$

Because the swelling in at least one footpad among four causes the thickness of the hind footpad to be increased by 15%, it was declared arthritic.

Lim et al. (2017) experimented with their study based on the CIA rat model of RA. They used Lewis rats (female) weighing in the range of 1.75–2 kg and caged them in digital smart houses. A 12 h light–dark cycle was maintained with unrestricted food and water for all the animals. Initially, rats were given anesthesia using isoflurane, and 2 mg/mL of type II collagen mixed with IFA was administered intradermally at the hind paw. After 7 days post-induction duration, a booster dose was injected into the rats. IFA was injected into the control animals.

Vium digital houses are equipped with sensors which stream the rat motion and its activities 24×7. The real-time data about the animal motion, humidity, respiratory rate, and temperature were collected, displayed, and analyzed by the vium digital platform. The animal activities are captured and monitored by using visible and infrared video cameras at the frame rate of 24 frames/s. The digital arthritis index (DAI) was calculated from the data analyzed in the experimental study. The video data captured are converted into continuous time frames as still images. It was observed from the video data that when the arthritis started developing in the rats, their movement is eventually restricted, which, in turn, causes a reduced speed of motion. The DAI is calculated as the ratio of accumulated prominent features to the baseline features. It was generated from the captured video images based on the maximal speed of motion during the light–dark cycle for all the rats. DAI is inversely proportional to the overall activity of the rats and exhibits a negative correlation. Because increased DAI depicts the decreased accumulated salient feature on a particular day. During the entire period of the study, the disease progressions were quantitatively evaluated based on the parameters such as body weights, limb circumference measurements, ankle joint sizes, and clinical arthritis score. The digital calipers were used for circumference measurement of ankle joints during the zeroth day, booster day, and on each subsequent day. The total arthritis score was calculated based on the scoring adopted for each hind paw for the appearance of inflammation and swelling as follows: 0 represents normal, 1 indicates mild redness and swelling of joints, 2 specifies moderate redness and swelling, 3 indicates the occurrence of severe redness and swelling, and 4 depicts the limb inflamed maximally in all the multiple joints. The authors obtained the DAI ankle joint measurements and clinical arthritis scores based on the following criteria: (i) incidence of disease, (ii) onset of disease, and (iii) severity of disease. The rats were considered to be arthritic only when the ankle joint, DAI, and clinical arthritis scores were greater than the threshold. The threshold for ankle joint was found to be 0.63 cm, cut off value for DAI was 5, and arthritis score's threshold value was set at 4. Dai et al. (2018) demonstrated the study on the CIA rat model and tested the efficacy and anti-arthritis properties of curcumin for the treatment of RA. They used Wistar rats (male) aged 7–8 weeks with an average weighing of 1 kg/rat. The total numbers of rats were categorized into two groups: Group 1: 10 rats (control group) and Group 2: 30 rats (CIA-induced RA). The authors liquefied the type II collagen in 0.05 M acetic acid and kept it overnight at a temperature of 4°C. Then, the liquefied type-II collagen was amalgamated with IFA in a 1:1 ratio and yielded 1 mg/mL concentration for injection in the Wistar rats. A booster injection consisting of collagen amalgamation of 300 μL and 200 μL was induced in the rat subcutaneously on day 1 and day 8, respectively. Three independent researchers performed arthritis scoring in each hind paw of Wistar rats and recorded the value as an average score. The severity of arthritis is scored in the range of 0–4, where 0 indicates the absence of swelling, 1 represents the presence of mild swelling in the ankle and foot, 2 shows the presence of moderate inflammation and swelling in ankle and foot, 3 represents the spread of inflammation and swelling in entire foot, and 4 indicates severe inflammation leads to loss of movement. The arthritic index was obtained as a sum of score of 16 per rat. An arthritic index of above 6 points is considered to be confirmed arthritis in rats.

4.1.2 MONO-ARTHRITIS MODEL

Brenner et al. (2005) conducted the experiments using 8–12-week-old female DA, ACI, and F344 based on the systemic autoimmune arthritis model and the mono-arthritis model. The mono-arthritis model is a localized model which begins after 2–6 h upon the onset of induction and the correlation studies were performed between the arthritis and thermal imaging parameters. They injected a 40 μL composition of CFA mixed with normal saline in the left ankle and compared it with the right ankle saline injection in each of the study rats. For the scoring parameter, they measured bilateral ankle diameters where a digital caliper to the nearest 100th mm was used to measure the diameter of an ankle. To assess the systemic autoimmune arthritis, the ankle and midfoot joints were considered, where the severity of arthritis was scored ranging from 0 to 4, where 0 signifies no swelling and 4 signifies severe swelling and non-weight bearing. Pain-related posture was scored at the animal's hind limb based on the rat's pain-related behavioral scale. The scoring was performed in the range of 0–5 as follows: 0 shows normal, 1 indicates the bending toes, 2 represents the inversion of paw, 3 shows the limited weight posture, 4 shows the non-weighted posture, and 5 represents the proof of contact with the hind limb. DA rats were injected with 150 µL of pristane dose which induced arthritis (pristane-induced arthritis [PIA]) beginning after 14 days of induction. They reported that the prevalence of arthritis in DA rats is extremely prone to PIA. Thus, the PIA method was used to determine if the thermal changes occurred prior to the onset can reflect pre-clinical synovitis which is further used for predictions of histological changes. Then, the rats were scored daily between 0 and 16 days using a well-established scoring system. Their arthritis-scoring system includes scoring based on the following categories: a) region-wise as wrist, mid-fore paw, ankle, and midfoot joint and b) based on the occurrence of arthritis in a number of joints, at least each of the three joints. The total scoring was obtained by summation of all individual joints score.

For imaging the rats using thermal imaging, in a sealed acrylic chamber, the rats were induced with anesthesia with 5% isoflurane in oxygen. After this, the rats were transferred to an IR imaging chamber with 2–3% isoflurane maintained with oxygen. The body temperatures of the rats were maintained at 37°C. The legs were placed in flexion and external rotation by exposing the medial part of each ankle. A total of 10 min was utilized for imaging each rat where the image was captured every 15 s. An average of 16 frames of the video were clubbed together to form one clear still image of the area of interest. The mean temperature within the prescribed ROI was measured and recorded for each image. The thermal image was acquired at 0 h, 6 h, 12 h, 24 h until day 7 after the injection of intra-articular saline. Further, the thermal images are acquired from day 0 to day 16 in the case of the PIA method.

Yu et al. (2006) purchased Sprague–Dawley (SD) (male) rats weighing 1.4–1.6 kg from Shanghai BK Experimental Animal Center, China. Under regular laboratory settings, all rats were acclimatized. In the breeding room, the lighting was turned on for 12 h and the temperature in the room was 24°C. The Committee of Laboratory Animals of Anhui Medical University, China, gave its approval to the experimental procedures. A total of 12 SD rats were split into two groups and 6 animals served as the model, while the other 6 were given a hypodermic injection of normal saline in the posterior paw (left side) as a control. A single hypodermic injection of 100 µL of FCA into the hind paw (left side) was used to develop arthritis in rats. 1 mg of heat-inactive Bacillus Calmette–Guérin (BCG) combined with 0.1 cc sterile mineral oil is used for the preparation of FCA; the control animals were given a saline injection in their hind paws (left). On days 0, 10, 12, 14, 16, and 18 after the injection, the right posterior paw's volume was calculated using a volume meter. On the same days, the rats' body weights were assessed using an animal scale.

After FCA was injected into the left hind paws hypodermically, the right (uninjected) hind paws of AIA rats swelled for five days and then developed progressively. On day 16, the edematous response reached its peak. Normal rats, on the other hand, had very little edema and their right hind paw volume rose very little. Adjuvant-induced arthritic rats acquired much less weight than control rats. Following FCA injection, individual AIA rats gained a mean of 6 g over 10 days, with little change in body weight between days 10 and 14; it continued to grow in the next 14 days. During the first 10 days, control rats attained a weight gain of 51 g. Inspite of decreased weight gain in control rats from day 10 to day 18, the weight of control rats is significantly higher than the AIA rats. The animals' soft tissue ($p < 0.01$), bone detail ($p < 0.01$), and lesions ($p < 0.001$) were found to be significantly different between the X-ray phase-contrast imaging (XPCI) and the other two modalities, although there was no notable variance between absorption-contrast imaging and traditional soft X-ray radiography. With the scoring system, interobserver agreement was good (K = 0.618). On the tenth day after receiving a hypodermic injection of FCA, no bone alterations in the right hind paw's metatarsophalangeal joint were observed, although soft tissue edema was noticeable. By day 12, the reticular facet had a few cyst-like bone erosions with a diameter of about 0.2 mm. Bone erosions 0.1–0.2 mm wide were found in the area vicinity to the articular facet. Osteoporosis was discovered, and the soft tissue enlargement around it became damaged. By day 14, XPCI revealed that the bone erosions had increased, especially in the edge area of the subarticular facet. On day 16, there was no significant change in the bone erosions, and found to be in the same condition as in day-14 pictures; the joint space appeared to be broader than normal, and soft tissue edema was noticeable. On the 18th day, periosteal new bone growth was visible along the paracortex. Also, it appears as diminished bone degradation in the articular facet and the space; visible hyperostosis and reduced soft tissue inflammation were observed. All of these changes were visible on XPCI, but cartilage and synovium were not. The images of the rat hind paws on traditional X-ray depicted cortical bone, joint space, and medullary chamber of bone. Synchrotron radiation X-ray absorption-contrast imaging was clear compared to the traditional X-ray. In these two imaging modalities, however, we're unable to discern between the bone alterations and enlargement of surrounding soft tissue that was visible in XPCI. The AIA rats' synovial tissue grew, containing a high amount of fibroblasts and hyperplastic lining cells. Inflammatory synovial tissue generated new blood vessels in the meanwhile. Synovial cells showed fatty alterations as well. Following matrix deterioration, new blood vessels developed around the articular cartilage and subchondral bone, were eliminated. In four of the six rats studied, periosteal new bone growth was discovered.

Almarestani et al. (2011) conducted the experiments on Sprague–Dawley rats (male) to study the pain-related behavior. 150 µL of CFA was injected into the right hind paw. They divided the rats into two groups as follows: Group 1 – saline injection and Group 2 – CFA-induced. An initial period of acclimatization of rats was followed before the testing process. They tested the plantar surface of the hind paw with the following: (i) mechanical hyperalgesia, (ii) cold nociception, and (iii) heat allodynic. In the case of mechanical hyperalgesia, they tested the rats by giving mechanical stimuli using von-Frey filaments (Tal and Bennett. 1994). The rats were placed on an elevated wire mesh platform that permits access to all four paws. Initially, a 4 g von-Frey filament is applied to the plantar surface of the hind paws (left-side paw first proceeded by the right). Then, a 15 g filament was applied. Totally, 20 times, both the filaments were applied and withdrawal reflex was recorded. Slow withdrawal, biting, and movement of limb were taken as positive responses. In the case of cold nociception, acetone drop method was used to study the pain characteristics in rats (Choi et al., 1994). The animals were placed on a wire mesh floor. 50 µL acetone was taken in a 1 mL syringe and slowly, a drop of acetone was applied to the plantar surface of the hind paw three times with an interval of 5 minutes. The response of normal rats is minimal, but CFA-induced rats provided a good response to acetone application. A scoring method was followed ranging from 0 to 3 in which score 0 denotes no response, score 1 indicates quick flick, score 2 represents the continuous flick, and score 3 shows biting of the paw. After applying this cold stimulus, the rats were given a 20 s duration to respond to the stimulus.

In case of heat allodynic, the rats were kept on a glass floor and a light source is passed underneath the floor which exposed the heel region of rats (Hargreaves et al., 1988). The withdrawal reflex disrupted the reflected light from the heel onto the photocell. This made the light and timer automatically turn off. The baseline latencies were observed as 7 s with a maximum cut-off frequency of 15 s to overcome tissue damage.

Snekhalatha et al. (2012) performed the experiments on 7–8-week-old Wistar rats to study the thermal imaging characteristics and early prediction of arthritis. The authors injected 0.2 mL of CFA mixed with PBS in a 1:1 ratio into the right side forepaw and hind paw and kept the left side paws as control. After the 5th day, a booster injection is provided with 0.1 mL of emulsion to increase the severity of arthritis. Limb circumference measurements were performed in both the inflamed and non-inflamed paws of both limbs. A semi-quantitative scoring method is adopted to study the severity of arthritis on each subsequent day as follows: 0 indicates normal with no signs of arthritis, 1 represents mild swelling in the inflamed joints, 2 shows the moderate swelling and redness persist in more than one joints, 3 depicts the swelling and redness in two or more joints, and 4 represents severe arthritis in more than three joints, i.e., multiple joints (Figure 4.1). Then, the authors obtained the thermal imaging measurements from day 0 to each subsequent day for all the rats, followed by radiographs attained on the 25th day to measure arthritis parameters such as erosion, joint space narrowing, and joint space destruction.

Xia et al. demonstrated their experiment on male Wistar rats 3–4 weeks old. All the rats were acclimatized with maximum room temperature maintained at 25°C

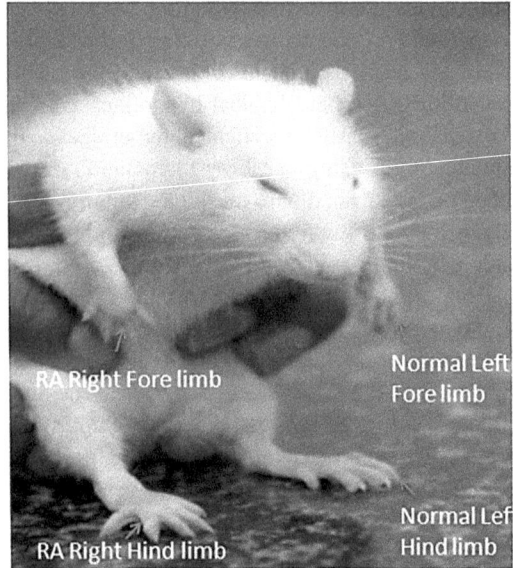

FIGURE 4.1 Induced RA in Wistar rat using CFA in which the picture depicts inflammation in the right forelimb and hind limb whereas it is normal in the case of left forelimb and hind limb.

of 40–50% humidity with access to good food and water. The rats were categorized into two groups with five rats per group.

- **Group 1**: CFA-induced in right hind paw
- **Group 2**: CFA-induced in left hind paw

0.25 mL of CFA was injected in both the right hind paw of Group 1 and the left hind paw of Group 2 (Xia et al. 2021). The contralateral paw of each group was kept as control. The control side was injected with normal saline. The pain tolerance test was conducted for each rat using a plantar test device. The rats were kept in a plastic container and placed on a glass floor. A beam of radiant heat is passed to the hind paw through the glass floor. Four trials were conducted for an interval of 10 minutes and baseline latencies were measured.

4.1.3 PRISTANE-INDUCED ARTHRITIS

Pristane is a non-antigenic chemical substance which induces delayed occurrence of arthritis in rats. CD^{4+} T-cells, heat shock protein 65, and lymphocytes cause chronic arthritis in rats. The chronic arthritis is onset from 60 to 200 days in the rodent by inducing intra-peritoneal injection of pristane. Hence, this chronic arthritis induced due to PIA resembles RA. The clinical characteristic of PIA involves joint swelling followed by pannus formation in affected joints and infiltration of polymorphonuclear cells. The arthritis development is observed by following a macroscopic scoring system which produces the scoring ranging from 0 to 4 in four limbs. PIA is a kind of autoimmune arthritis model in which the onset of the disease occurs four days after the induction. The authors used DA rats to detect the pre-clinical synovitis and histological findings. 150 μL of pristane (2,6,10,14-tetramethyl pentadecane) was injected at the base of the tail in DA rats. They followed an arthritis-scoring system to evaluate the individual joints in various regions such as wrist, mid-fore paw, ankle, and mid-foot joints according to scores ranging from 0 to 4. They obtained the maximum score at each extremity and total joint per rat as 20 and 80, respectively.

Tuncel et al. (2016) obtained 217 rats of ages ranging from 7 to 20 weeks. Rats in groups of five were caged together and were fed autoclaved food and *albidum* water and were subjected to 14 h light and 10 h dark cycles. The normal rats were also fed the same diet as the barrier rats. 100 μL of pristane was injected intradermally at the dorsal side of the tail in 8–11-week-old rats. Next, 100 μL IFA of oil-induced arthritis (OIA) and CIA were dissolved in 150 μL of 0.1 M acetic acid and were administered intradermally to the rats. For the scoring, 1–5 points were assigned to the inflamed muscle or toe of the affected ankle. If the deformation accompanied by erythema was not observed, then the scoring was eliminated. The occurrence of chronic arthritis is illustrated in the experimental rats with ≥ 5 mean score or at least ≥ 6 score for 2 days following 60 days after the immunization. To perform the comparison between late stages of PIA and OIA based on histopathological studies, after 130 days of post injection of pristane or IFA, the paws were dissected and decalcified with EDTA solution. Tartrate-resistant acid phosphatase (TRAP) staining was used to identify osteoclasts in paraffin-embedded tissue sections of hind paws. To visualize the calcified tissue, the paws were imaged with microcomputed tomography. Then, 3D volumes of the MicroCT images were segmented based on isotropic voxel of size 9 μm using open VMS by SCANCO.

For the treatment, they administered many compositions such as Etanercept, Fingolimod, Cyclosporine, Methotrexate, and Dexamethasone orally or i.p. Also, the control rats were administered with an equal volume of ethanol (50%)/PBS vehicles. The Mann–Whitney U test was performed for distinguishing the two non-normally distributed data which include the arthritis score, day of onset of diseases, weight changes in the rats, and day of maximum arthritis. They used Kruskal–Wallis test for multiple group comparison and reported their significance value. Linear regression was performed to find the correlation between acute and chronic arthritis. Moreover, they also performed coefficient of variation (CV) to obtain the arthritis scores between immunization groups in study animals.

4.1.4 ADJUVANT-INDUCED ARTHRITIS

It was observed from the findings that CFA administration in specific strains of rats causes arthritis which has been proved in the AIA model (Pearson 1956). In the year 1988, Van eden et al. observed that AIA is formed due to the cross-reaction of mycobacterium with joint-specific antigens. Later, the researchers predicted that non-antigenic components such as IFA and muramyl dipeptide can cause arthritis in rats. The major difference between AIA and CIA is that resistance is pre-dominant in AIA models.

4.1.5 ANTIGEN-INDUCED ARTHRITIS

Antigen-induced arthritis is a kind of mono-articular disease which affects only the injected joints. This type of arthritis is induced in knee joints of rats by means of injecting the protein antigen into intra-articular joints. The protein antigen is comprised of methylated bovine serum albumin. The histological results of antigen-induced arthritis closely resemble RA and depict synovial hyperplasia, pannus formation, cartilage erosions, and lymphoid follicles. Certainly, frequent antigen injection in rats causes ectopic lymphoid structures identical to those observed in RA patients.

4.1.6 PROTEOGLYCAN (PG)-INDUCED ARTHRITIS

PG-induced arthritis is a cartilage-specific PG protein called aggrecan. Its biological characteristics are similar to collagen arthritis. This aggrecan is used as an antigen for the induction of arthritis in rodents and mice. This PG aggrecan has been proclaimed to cause erosive polyarthritis in mice. The interaction between T-cells and B-cells exactly correlated to PG-induced arthritis (Banerjee et al., 1992). These T-cells and B-cells are considered pre-dominant immune cells in the origin of arthritis. Agglomeration of PG-specific antigen IgG in the inflamed joints and cartilage causes cartilage loss (Mikecz et al., 1990). The evolution of arthritis and its progression is mainly caused by the interaction of IgG immune complexes and Fc receptors for IgG (Wooley et al., 1992). A 12–14-week-old female mouse is arthritis-induced using 150 µg PG (obtained from human arthritic cartilage patients) injected intraperitoneally. A booster dose of 100 µg PG was given to the mouse with the first booster dose at third week and the second booster dose at sixth week time intervals. Symptoms of arthritis such as edema and ankylosis were observed and scored in the range of 0–4 for every two-week intervals, and the scoring process was carried out nine times until the 18th week and the arthritic index based on scoring was obtained. Then, at the 18th week's end, histopathological studies were carried out by dissecting the hind ankles of mouse and put in a container containing formalin, embedded in paraffin, and finally, the sample is stained with hematoxylin and eosin. After that, the microscopic observation revealed the effect of PG-induced arthritis as pannus formation, cartilage erosions, and mononuclear cell infiltration in the histological report. The mice model of PG arthritis closely resembles the clinical features of human RA which is evident from the radiographic and histopathological studies.

4.1.7 CARRAGEENIN-INDUCED ARTHRITIS

Sanchez et al. (2008) conducted the study in a rat model using carrageenin-induced arthritis. To induce carrageenin, male Sprague–Dawley rats of 125–150 g weights were used. Prior to administration of carrageenin, an anesthetic agent named isoflurane is injected into the rats. 2 mg of carrageenin is dissolved in 0.1 mL saline and the solution is injected into the right hind paw of the rat. The left hind paw was injected with 0.9% saline. After 4 h post administration of carrageenin, thermal images were acquired to measure the skin surface temperature and paw swellings were measured.

Calkosinski et al. (2015) selected 16 female rats having body mass of 150–160 g and were 10 weeks old from the inbred *Buffalo* strain rats. The age and mass of the study rats were maintained the same as some inflammatory factors highly depend on the age, sex, and the strain of the study rats. The area of the right intercostal space (near the elbow joint) was chemically depleted for 2 h before the measurement of temperature in the first group. 0.9% of NaCl solution was injected as a control at 2–3 mm depth in all the study rats which did not cause any change in temperature in the rats' bodies. They also divided their study animal into four groups containing four female rats in each group. In Group 1, they evoked pleural inflammatory reaction by administrating 0.15 mL of carrageenin solution between 5 and 6 intercostal space which is the region for temperature measurement. In Group 2, they evoked an inflammatory reaction by administrating 0.15 mL of 1% carrageenin solution in the lower lip from the side of the oral vestibule. In Group 3, they administrated 0.15 mL of 1% of carrageenin in the foot region of lower extremities of the right limb (paw pad), and in the final group (Group 4), they evoked the inflammatory reaction in the foot of the lower left limb by administrating 0.15 mL of 1% carrageenin. They captured the thermal images of regions such as lower lip, pleura, and left and right paws using Therma CAM 550 Researcher Pro 2.8 software. The thickness of paws was measured in inflamed limbs and control using a digital caliper. The temperature was obtained in specific groups and the inflammatory reaction within various body regions was performed at different time intervals. The measurement of temperature was performed in control before the occurrence of inflammatory reactions. After the administration of carrageenin, the measurement was made at different time intervals to evaluate the characteristics of inflammation. They also administered 100 mg/kg pharmacological euthanasia intraperitoneally to the animals. They also performed statistical analysis where they reported that temperature difference existed between the analyzed site and normal distribution was verified using Shapiro–Wilk test. The differences in significance between regions with inflammatory reactions and without inflammatory reactions were verified using Kruskal–Wallis ANOVA and standard ANOVA techniques.

4.2   THERMAL IMAGE ACQUISITION

A portable thermal camera was used for imaging the forelimb and hind limb of inflamed and control regions. A constant-temperature-controlled room was used for acquiring the thermal images of Wistar rats. The rats are placed in a supine position. A wooden cardboard with a slit of size 10 × 10 cm was placed as partition in both the limbs. The limbs are exposed to the outside through the slit hole.

The images were acquired in the posterior–anterior view only in the exposed portion of the limbs. The thermal images were acquired by positioning the camera at a distance of 1 m and the images were obtained from day 1 to day 30 after being CFA-induced (Snekhalatha et al., 2012). FLIR Version 1.2 was used to analyze the temperature profile of each rat on day to day basis.

Sanchez et al. conducted their experiment in a temperature-controlled room with an ambient temperature of 70.7 ± 2°F. A Thermoview Ti30 portable thermal camera was used to acquire the thermal images of hind paws. The camera used 120×160 pixel resolution of uncooled microbolometer detection system with the minimum distance focussed at 30 cm. Prior to image acquisition steps, the rats were anesthetized with isoflurane. After being anesthetized, the rats were placed in supine position with the hind paws lifted and positioned for imaging. A black velvet cloth is kept for background illumination. The thermal image is acquired by fixing the camera 30–38cm away from the hind limbs. The authors used inbuilt thermoview software to analyze the thermal image and displayed the average temperature from the required ROI. They captured the digital images to visualize the inflammation in the hind paw (right side) and the non-inflamed paw (left side).

4.3   THERMAL IMAGE ANALYSIS

After the thermal acquisition process, the thermal images were analyzed for the measurement of their skin temperature on each subsequent day. The thermal images acquired on days 0–10 were measured with skin temperature parameters using FLIR Version 2.0 as depicted in Table 4.1. Among the temperatures

TABLE 4.1
Day-Wise Thermal Profile of Small Animal Study

Day	Left Forelimb (Normal) (°C) n = 15	Right Forelimb (Arthritis-Induced) (°C) n = 15	Statistical Significance (p)	Left hind Limb (Normal) (°C) n = 15	Right hind Limb (Arthritis-Induced) (°C) n = 15	Statistical Significance (p)
0	34.30 ± 0.4	34.30 ± 0.9	0.8	34.33 ± 0.3	34.33 ± 0.6	0.7
1	34.30 ± 0.4	34.30 ± 0.8	0.9	34.33 ± 0.4	34.33 ± 0.8	0.8
2	34.33 ± 0.3	34.66 ± 0.9	0.06	34.35 ± 0.4	34.66 ± 0.6	0.05
3	34.63 ± 0.5	34.84 ± 0.7	0.05	34.65 ± 0.3	34.88 ± 0.7	0.02
4	34.66 ± 0.5	34.95 ± 0.9	0.002	34.66 ± 0.4	35.05 ± 0.6	0.0001
5	34.68 ± 0.4	35.05 ± 0.8	0.001	34.69 ± 0.3	35.25 ± 0.6	0.0002
6	34.69 ± 0.3	35.15 ± 0.6	0.001	34.70 ± 0.3	35.30 ± 0.7	0.0004
7	34.70 ± 0.4	35.20 ± 0.6	0.003	34.70 ± 0.4	35.40 ± 0.7	0.0004
8	34.70 ± 0.4	35.35 ± 0.7	0.002	34.72 ± 0.5	35.45 ± 0.6	$<10^{-05}$
9	34.72 ± 0.5	35.50 ± 0.9	0.002	34.75 ± 0.6	35.55 ± 0.7	$<10^{-05}$
10	34.80 ± 0.5	35.89 ± 0.9	0.002	34.75 ± 0.5	35.70 ± 0.6	$<10^{-04}$

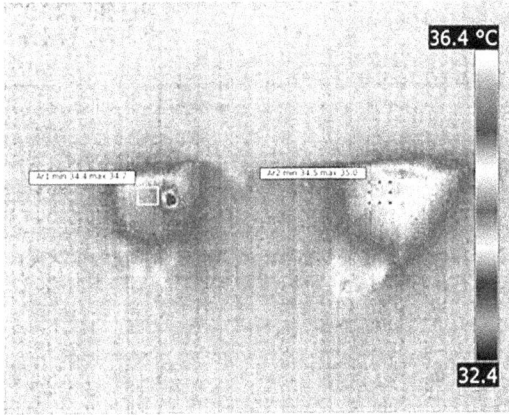

FIGURE 4.2 Thermal image acquired for Wistar rat on day 0.

measured on each day, a higher temperature difference of 1.09°C was observed between the normal and arthritis-induced forelimb on the tenth day. Similarly, a temperature difference of 0.95 was attained between the normal and arthritis-induced hind limb. A rectangular ROI is fixed on the image in both normal and inflamed limbs and temperatures were measured. The minimum and maximum temperatures were displayed in the required ROI, from which the mean temperature was calculated. Figure 4.2 shows the thermal image of Wistar rat acquired on the zeroeth day in which temperature measurements were made. Figure 4.3 denotes the skin temperature measured on the first day in both forelimb and hind limb of normal and arthritis-induced joints. Figure 4.4 indicates the skin temperature measured on the fourth day in both forelimb and hind limb of normal

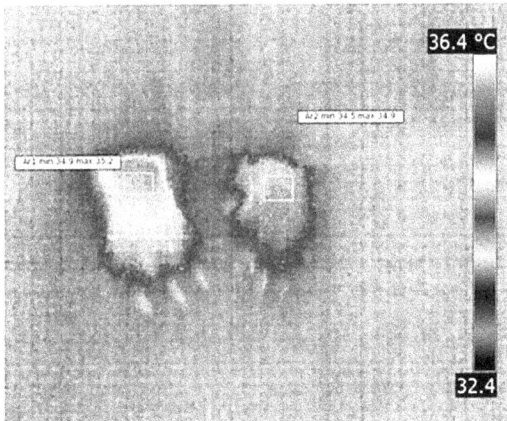

FIGURE 4.3 Temperature measurement performed on day 1 on inflamed and normal limb.

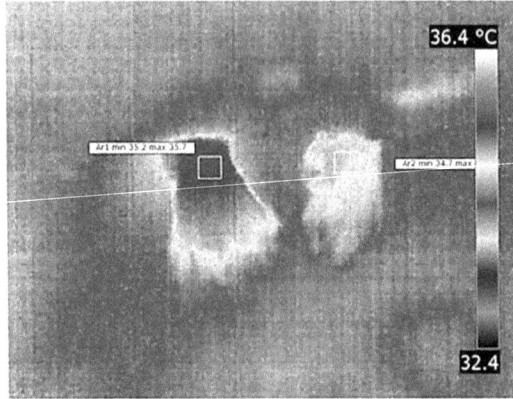

FIGURE 4.4 Temperature measurement performed on day 4 on inflamed and normal limb.

and arthritis-induced joints. Figure 4.5 depicts the temperature measurement performed on the tenth day on inflamed and normal limbs.

According to Jasemian et al. (2011), to ensure that the general temperature rise in the rat's body due to collagen immunization did not alter the recorded difference in temperature rise in the hind feet, temperatures on the hind feet (metatarsal and tarsus) were normalized with the shaved reference area on the back. The mean temperatures in each rat's hind foot (left: t left, right: t right) were normalized to the back temperatures (t back), and the temperature index (TI) was determined (TI = t left / t back + t right/t back). In the undamaged group, the normal hind foot temperature was roughly 26.55°C (TI 1.5), whereas hind feet had a TI > 1.5,

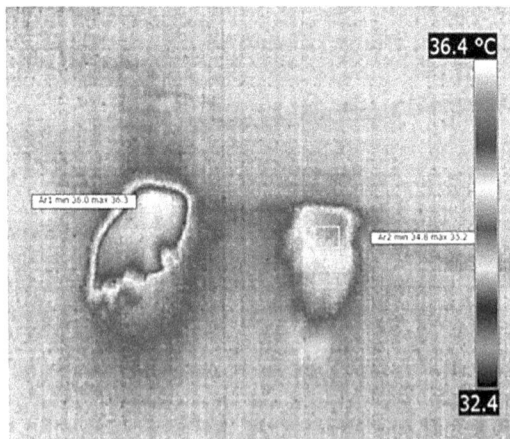

FIGURE 4.5 Temperature measurement performed on day 10 on inflamed and normal limb.

corresponding to a temperature of 27°C. The average rat back temperature was 36 ± 2°C. As expected, the rats in the vehicle group developed arthritis while dexamethasone-treated rats did not develop symptoms or developed only mild symptoms. The authors obtained no difference in clinical score between groups before day 17, whereas a significant increase in temperature on day 9 was observed on comparing vehicle versus dexamethasone-treated rat (difference of mean = 0.118, p < 0.011). The temperature index of 0.118 corresponds to 0.95°C.

Nosrati et al. used utilized thermographic imaging, a camera that is used to transform infrared light into an image that shows temperature changes throughout an object or area (2). Thermal pictures were captured in this investigation utilizing a FLIR ONE camera connected to an iPhone (model 6). FLIR ONE has two cameras: a Lepton™ thermal camera and a VGA light camera whose resolutions are 80 × 60 pixels and 1440 × 1080 pixels, respectively. This camera is capable of capturing both digital images and thermal images and fusing them using multispectral dynamic imaging (MSX) to attain the infrared images of improved characteristic and resolution. For picture visualization and analysis, FLIR Version 5.1 was utilized. A range of calibration methods is used to estimate the FLIR ONE camera's ability. Then, the authors went over the picture acquisition and processing method for the data.

The FLIR ONE camera used in their study had a temperature span of −20–120°C and can detect temperature changes as tiny as 0.1°C. A standardization check was carried out using boiling water, melting ice, and a person's tear duct with known temperature to test the instrument's performance before usage. To determine the device's precision in the temperature span of 35–40°C, it was tested against a black reference material with established emissivity at different temperatures. The coefficient of determination (R^2) was used to prove that the linear regression connecting the black body and IR temperature data explained the fluctuation. The temperature reading was taken three times and the mean SD was calculated. From a 20 cm distance, the camera was set horizontally parallel to the ground, focussing on hot water and melting ice. The distance and, as a result, the field of view (FOV) were retained persistent (approximately 20 cm) while focussing the attention toward the object and limiting the surroundings to picture the tear duct and black material.

Infrared cameras capture thermal data and display it as a color map. Different color palettes are provided, each of which assigns a specific color to the temperature readout readings, allowing visual interpretation of a thermal image. Both the camera and the PC processing software can be used for color assignment. FLIR Tools software was used to examine the thermographic pictures that were collected. The "iron" palette was chosen, with black symbolizing the coldest places, blue and magenta representing moderately warm regions, red, orange, and yellow representing the middle span of temperatures, and white representing the hottest region. To construct the temperature index, ROIs were generated and used to draw out temperature specifications. Elliptical ROIs were kept over the mouse's anterior and back paws, as well as three locations on its reverse side (reference ROIs). The ROIs were of the same size throughout the research. To compute the TI of each paw, the average skin surface temperatures (T) in individual ROI in various regions such as left forelimb, right forelimb, hind limb left, and hind limb right were standardized to the average reference temperatures on the backs (T_{back}) of each rat. In addition, for an individual mouse at an individual time point, the total temperature index (TI total) was determined as the sum of the wrist joint TIs and the ankle joint TIs of forepaws and hind paws, respectively.

According to Calkosinski et al., thermal image analysis included interactive bifurcation of the examined sections of body surface area and analysis of quartiles (minimum and maximum temperatures, median, and standard deviation of temperatures using FLIR software. A caliper was used to measure the paw thickness in the studied leg (with swelling reaction) and the control. Temperature measurements were taken at the following times in certain classes in which an experimental swelling reaction was elicited in different body parts: (1) maintain quantification before induction of inflammatory response, $t0$, (2) computation 10 minutes next to carrageenin conduction, $t_{0.2}$, (3) estimation at 1 h next to carrageenin conduction, t_1, (4) estimation at 2 h next to carrageenin conduction, t_2, (5) evaluation at 4 h next to carrageenin conduction, t_4, (6) measurement at 6 h next to carrageenin conduction, t_6, (7) assessment at 12 h next to carrageenin conduction, t_{12}, (8) calculation at 24 h next to inducing inflammatory reaction, t_{24}, (9) quantification at 48 h next to inducing inflammatory reaction, t_{48}, (10) estimation at 72 h next to inducing inflammatory reaction, t_{72}, and (11) evaluation at 96 h next to inducing inflammatory reaction, t_{96}

A literature survey of applications of thermal imaging modality in various models of arthritis in small animal model study and their key observations are depicted in Table 4.2.

4.4 HISTOPATHOLOGY EVALUATION

4.4.1 HISTOPATHOLOGICAL ANALYSIS OF SYNOVIAL FLUID

A specimen of the RA affected joint is obtained through closed needle biopsy or arthroscopic guided biopsy. Then, it is stained with hematoxylin and eosin. The histological evaluation of synovial tissue inflammation in RA confirms the hyperplasia in the lining layer, cellular infiltration with lymphocytes, macrophages, granulomas, pannus formation, and plasma cells (Snekhalatha et al., 2012). In the Wistar rat-RA model, to confirm the pathological conditions, a histology study on rats was carried out by sacrificing on the 30th day and dissecting their forepaw and hind paw.

The ankle joints are dissected and placed in a Surgipath decalcifier which contains 10% formalin for about 1 week duration. HCL and 0.1 M EDTA solution were used for the paws' decalcification process. The dissected ankle joints were cut into two equal halves and encapsulated in paraffin and stained in hematoxylin eosin. The histological scoring is implemented for the skeletal sites as follows: tibio-talar, talus-calcaneal, mid-foot joints, and mid-finger joints. In the Wistar rat-RA model, the published histological scoring system is used to measure the

TABLE 4.2

An Elaborate Survey on Small Animal Model of Arthritis Based on Thermal Imaging

Author and Year	Animal	Induction	Imaging Method	Inference
Brenner et al. (2005)	Female DA rats (8–12 weeks)	Mono-arthritis model – CFA PIA-induced model-pristane	Thermal imaging	• Rats developed arthritis after 6 h injection of CFA. • Arthritis severity score correlated with thermal imaging in ankle joints.
Sanchez et al. (2008)	Female Dark Agouti rats weighing 110–125 g	Carrageenin-induced arthritis collagen-induced arthritis model	Thermal imaging	• Arthritis developed after 2 h post induction of carrageenin. • Predominant increase in average temperature difference between inflamed paw and control.
Jasemian et al. (2011)	Adult female Lewis rats weighing 160–187 g	Porcine type II collagen-induced arthritis	Thermal imaging	• CIA model produced elevated temperature in the hind feet region which has been treated with dexamethasone. • A significant positive correlation was observed between the hind foot temperature and clinical scores.
Almarestani et al. (2011)	Male Sprague–Dawley rats	150 μL CFA	Thermal imaging	• Arthritis signs started in rats after 2 weeks of post induction. • Developed inflammation in soft tissue and polyarthritis.
Calkosinski et al. (2015)	16 female rats	Carrageenin-induced arthritis	Thermal imaging	• Carrageenin injection initially causes a decrease in temperature but reached a maximum temperature on the third day of injection. • Difference in temperature between the inflamed and control is 1.3–1.5°C
Tuncel et al. (2016)	8–11-week rat	150 μL pristane-induced and 100 μL IFA-induced	Thermal imaging	• 99.6% of rats developed arthritis after the administration of Pristane. • Has great incidence and reproducibility. • More suitable than AIA.
Nosratiz et al. (2020)	7-week-old DBA1/J mice	CFA by type-II collagen emulsified in IFA	Thermal imaging	• Positive correlation was predicted between the temperature measurements and paw thickness. • Quantify the degree of inflammation in a reproducible method.

following variables semi-quantitatively with scores ranging from 0 to 3: (i) enlargement of the synovial lining cell layer, (ii) synovial hyperplasia, (iii) inflammatory infiltrate, (iv) synovial vascularity, (v) cartilage erosions, (vi) bone erosions, and (vii) pannus formation (Bresnihan 2003, Krenn et al., 2006, Koizumi et al., 1999). A score of 0 indicates the absence of a particular variable and a score of 1 illustrates the severity of a particular variable. In the evaluation of histopathological studies in the Wistar rat-RA model, the mean histological score for inflammatory infiltrate, pannus formation, and synovial hyperplasia were 2.8 ± 0.4, 2 ± 0.7, and 2.8 ± 0.8, respectively.

Section studied from the synovial joint shows mild synovial hyperplasia, depicted as long arrow in Figure 4.6a with adjacent subchondral bone layer represented as short arrow in Figure 4.6a. The synovial layer shows a mild enlargement of the synovial lining cell layer, synovial vascularity appears unremarkable, indicated as short arrow in Figure 4.6b, and scattered mononuclear inflammatory infiltration comprising predominantly lymphocytes and some eosinophils is depicted as long arrow in Figure 4.6b.

Section studied from the synovial joint shows synovial hyperplasia as mentioned as long arrow in Figure 4.7a and with adjacent subchondral bone layer marked as short arrow in Figure 4.7a. The synovial layer shows the enlargement of the synovial lining cell layer, and increased synovial vascularity is depicted as short arrow in Figure 4.7b, and mononuclear inflammatory infiltration comprising predominantly lymphocytes and plasma cells is depicted as long arrow in Figure 4.7b.

Section studied from the synovial joint shows focal synovial hyperplasia as mentioned in Figure 4.8a with adjacent subchondral bone layer given as short arrow in Figure 4.8a. The synovial layer shows the enlargement of the synovial lining cell layer and increased synovial vascularity, giant cells as depicted as short arrow in Figure 4.8b, and mononuclear inflammatory infiltration comprising predominantly lymphocytes and histiocytes is marked as long arrow in Figure 4.8b.

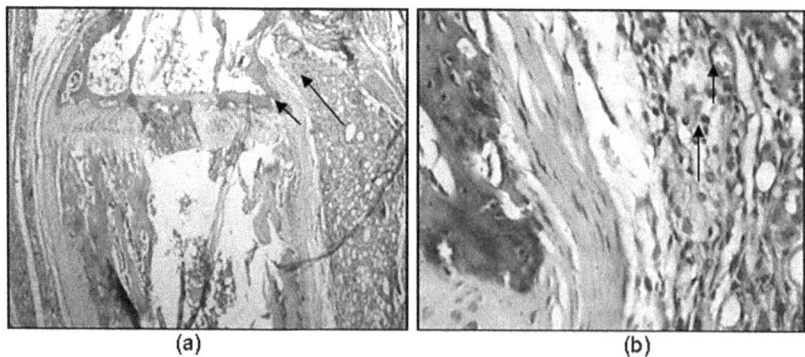

(a)   (b)

FIGURE 4.6 (a) Histopathological image depicting mild synovial hyperplasia in the synovial joint imaged at 50× magnification. (b) Histopathological image representing predominant lymphocytes in the synovial joint imaged at 400× magnification.

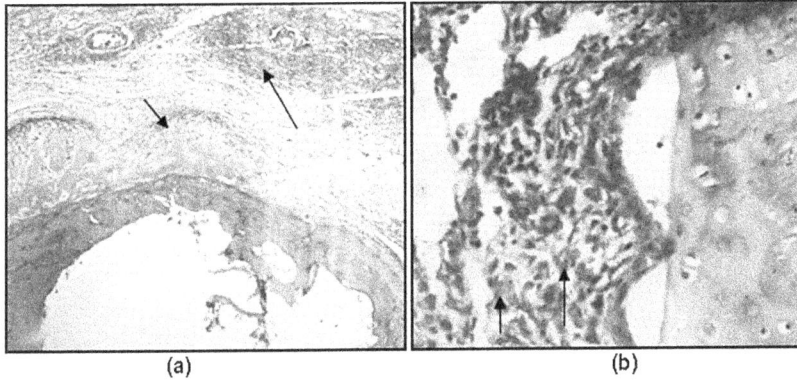

FIGURE 4.7   (a) Synovial hyperplasia in the synovial joint depicting adjacent subchondral bone layer imaged at 50× magnification. (b) Mononuclear infiltration and plasma cells imaged at 400× magnification.

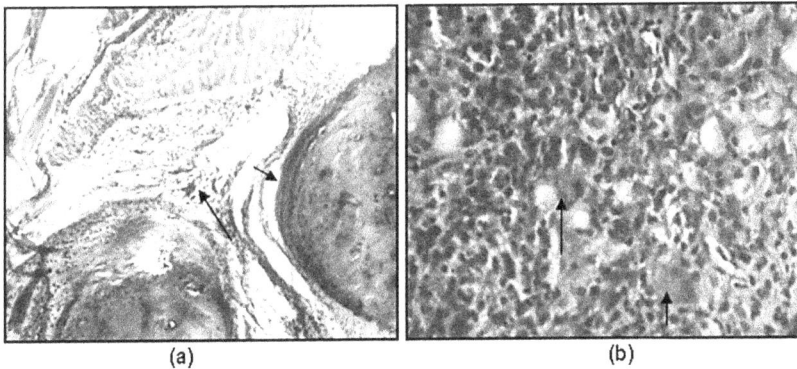

FIGURE 4.8   (a) Focal synovial hyperplasia in synovial joint imaged at 50× magnification. (b) Histiocytes in synovial joint imaged at 400× magnification.

Section studied from the synovial joint shows mild synovial hyperplasia marked as long arrow in Figure 4.9a with adjacent subchondral bone layer depicted as short arrow in Figure 4.9a. The synovial layer shows the focal enlargement of the synovial lining cell layer, synovial vascularity appears unremarkable, and mononuclear inflammatory infiltration comprising predominantly lymphocytes and few histiocytes is shown as long arrow in Figure 4.9b.

Figure 4.10 represents the histopathological image obtained for forelimb and hind limb of normal and diseased, respectively. Figure 4.10a indicates the normal hind limb depicting the dermis comprising fibrocollagenous stroma with adnexal structures, Figure 4.10b represents the diseased forelimb in which the dermis layer depicts the dense lympho-histiocytic infiltration, Figure 4.10c shows the diseased hindlimb representing the agglomeration of mononuclear inflammatory

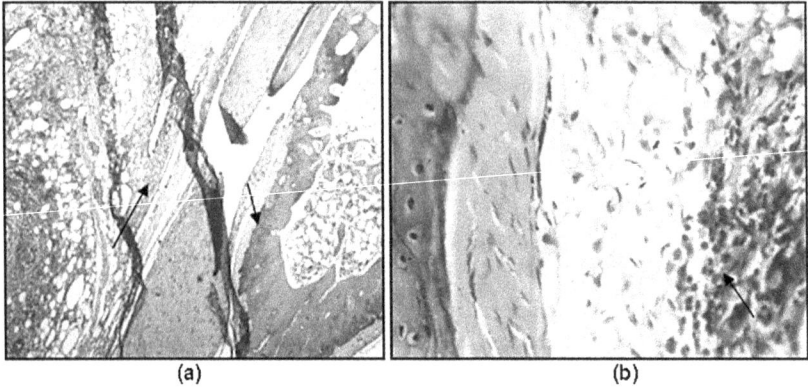

FIGURE 4.9 (a) Mild synovial hyperplasia imaged at 50× magnification. (b) Synovial vascularity showing the predominant lymphocytes and a few histiocytes images imaged at 400× magnification (H&E, 400×).

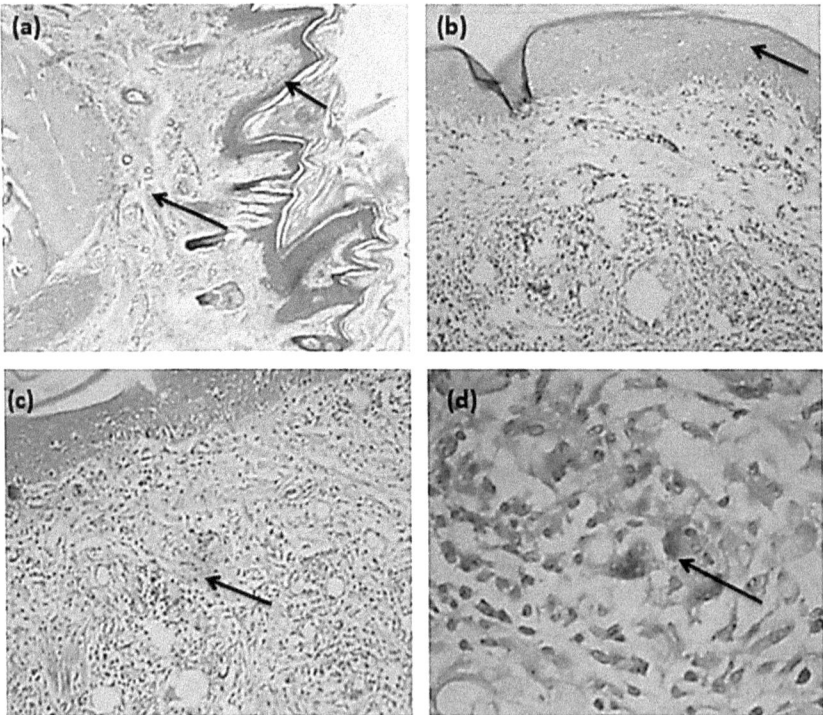

FIGURE 4.10 (a) Normal hind limb depicting the epidermis lined by stratified squamous epithelium and imaged at 40× magnification. (b) Diseased forelimb represents the dermis showing dense lympho-histiocytic infiltration imaged at 50× magnification. (c) Diseased hind limb depicting the aggregates of mononuclear inflammatory infiltration imaged at 50× magnification. (d) Diseased hind limb represents the giant cells imaged at 400× magnification.

infiltration, and Figure 4.10d depicts the presence of giant cells in the diseased hind limb imaged at 400× magnification.

For the histology and histologic scoring, the hind paws of the study rats were placed in formaldehyde (10%) at the end of the seventh day for mono-arthritis observation and for 16 days in the case of the PIA method (Brenner et al., 2006). The paws were then de-calcified with hydrochloric acid and 0.1 M EDTA solution. For the historical scoring, parameters such as inflammation in synovial joints, synovial hyperplasia, addendum of pannus formation, synovial fibrosis, synovial vascularity, and cartilage and bone erosion were studied and scored accordingly. The following criteria were graded histologically in the tibiotalar, talus-calcaneal, and midfoot joints. (a) Inflammation of the synovium. Five high-power magnification fields (HMFs) were used for invading the mononuclear cells and evaluated as follows: 0 indicates that normal, 1 indicates that something is lenient (1–10%), 2 indicates that something is medium (11–50%), and 3 indicates that something is extreme (51–100%). For the analysis, the mean of the five HMFs was employed. (b) Synovial hyperplasia is a type of synovial hyperplasia: 0 means no layer, 1 means gentle (5–10 layers), 2 means medium (11–20 layers), and 3 means extreme (>20 layers). (c) Pannus formation is extended based on the reader's impression: 0 indicates that the condition is absent: 1 specifies that it is light, 2 indicates that it is medium, and 3 specifies that it is extreme. (d) Fibrosis of the synovium was evaluated as follows: 0 indicates that something is missing, 1 indicates that something is mild (1–10%), 2 indicates that something is moderate (11–50%), and 3 indicates that something is severe (51–100%); (e) vascularity of the synovium (angiogenesis). The number of vessels in five HMFs of synovial tissue was counted, and the average was used for analysis, (f) erosion of cartilage. The percentage of the cartilage surface that has been degraded as follows: 0 indicates that something is missing, 1 indicates that something is mild (1–10%), 2 indicates that something is moderate (11–50%), and 3 indicates that something is severe (51–100%). (g) Erosion of the bones is graded as follows: 0 indicates no erosion, 1 indicates gentle erosion (seen exclusively at HMF), 2 indicates medium (complied with low magnification), and 3 indicates extreme (transcortical) erosion.

Chan et al., (2015), during the final stage of their experimental study, conducted histopathological studies by detaching the rat limbs and placing them in 10% buffered formalin; then, 10% of ethylene diamine tetra-acetic acid is used for decalcification and kept in paraffin and the sample is then dissected into sections and stained in hematoxylin and eosin. The entire morphology, infiltration of cells, and annihilation of tissues were seen in the microscopic digital image.

Lim et al. (2017) performed histological procedures in that tibiotarsal joints were dissected from the rat and cut into saggital hemi sections. The histological analysis was executed by a certified pathologist. The dissected joints were placed in 10% buffered formalin solution and kept in a Surgipath decalcifier. Then, the joint section was allowed to stain in hematoxylin and eosin and with toluidine blue to access the cartilage pathology. Histological scoring was given to all the joints to evaluate the cartilage damage, degree of inflammation, bone resorption, pannus formation, and periosteal bone formation. Then, the overall histological score is obtained as the sum of individual scores for an overall score of 25. The mean score was obtained based on individual calculations of right- and left hind joints and the average values of both joints were computed.

Finally, on day 18, all six AIA rats' uninjected hind paws (distal to the ankle joint), replete with muscles, tendons, and ligaments, were severed, preserved in 10% neutral buffered formalin for 48 h, and then decalcified for another 48 h (Yu et al., 2006). The specimens were sliced lengthwise and then dried in graded alcohol before being embedded in paraffin wax. Hematoxylin and eosin were used to stain sections of the bone that were 5 μm thick (H&E). One-way ANOVA was used to evaluate the changes in right hind paw edema and body weight. The weighted kappa test was used to examine the interobserver differences in the scoring system.

Dai et al. (2018) implemented histological studies on paw joints of the rat. They gutted the paw joints from the hind limbs of the rat. They removed the skin, superficial muscle, and tissue from the limbs and placed the sample in polyoxymethylene (4%) with the temperature maintained at 4°C. Then, for the decalcification process, the specimen was kept in 500 g of EDTA solution, 99.6 mL of HCL, and 3.46 L of deionized water for 35 days. For every five days, the EDTA solution was replaced with a new solution. The joints were drained with alcohol and instilled in paraffin. Finally, the joints were separated into slices of 4–6 μm thickness and fixed on a slide coated with a poly-L-lysine. To decrease the alcohol concentration and water, slides were immersed in xylene and dipped in PBS solution thrice and incubated in 3% of hydrogen peroxide for a period of 10 minutes duration and kept in citrate buffer for 20 minutes with a temperature maintained at 95°C. The sectioned tissue is incubated with primary antibody for 20 minutes and secondary antibody for 60 minutes and washed again with PBS solution. Then, the tissues were stained with DAB and examined underneath an optical microscope. Further, the digital histopathological images were acquired and analyzed for confirmation of pathological conditions of RA.

4.5 CONCLUSION

In conclusion, the thermal imaging measurements correlated with arthritis severity scores, radiological scores, and ankle diameters. It was found from the thermal profile of the Wistar rat-RA model that the temperature of both forelimb and hind limb gradually increased from day 4 onward and achieved a higher temperature on day 10. This study showed that thermography picked up induced arthritis in the small animal model at the early stages of the disease. Hence, thermal imaging can be used to detect the arthritis at a preclinical stage.

In measured mean average temperature (°C), the percentage difference between inflamed and control was found to be significantly ($p < 0.01$) higher by 3.03 and 2.74% at the RA forelimbs and hind limbs, respectively, than in the corresponding normal limbs. From the thermal profile of the Wistar rat-RA model, it was observed that the temperature of both forelimb and hind limb gradually increased from day 4 onward and achieved a higher temperature on day 10. Hence, thermal imaging can be used to detect the arthritis at a preclinical stage itself.

REFERENCES

Almarestani L, Fitzcharles MA, Bennett GJ, da-Silva AR. Imaging studies in Freund's complete adjuvant model of regional polyarthritis, a model suitable for the study of pain mechanisms, in the rat. *Arthritis & Rheumatism* 2011; 63(6): 1573–1581.

Banerjee S, Webber C, Poole AR. The induction of arthritis in mice by the cartilage proteoglycan aggrecan: Roles of CD4+ and CD8+ T cells. *Cell Immunol* 1992; 144:347–357.

Brenner M, Braun C, Oster M, Gulko PS. Thermal signature analysis as a novel method for evaluating of inflammatory arthritis activity. *Ann Rheum Dis* 2006; 65(3):306–311.

Brenner M, Meng HC, Yarlett NC, Griffiths MM, Remmers EF, Wilder RL, Gulko PS The non-MHC quantitative trait locus Cia 10 contains a major arthritis gene and regulates disease severity, pannus formation and joint damage. *Arthritis Rheumatism* 2005; 52(1): 322–332.

Bresnihan B Are synovial biopsies of diagnostic value?. *Arthritis Research and Therapy* 2003; 5:271–278.

Calkosinski I, DobrzyNski M RosiNczuk J, Dudek K, Chrószcz A, Fita K, Dymarek R. The use of infrared thermography as a rapid, quantitative, and non-invasive method for evaluation of inflammation response in different anatomical regions of rats. *BioMed Research International.* 2015, Article ID 972535.

Chan, Gray BD, Pak KY, Fong D. Non-invasive in vivo imaging of arthritis in a collagen-induced murine model with phosphatidylserine-binding near-infrared (NIR) dye. *Arthritis Research & Therapy* 2015; 17:50.

Choi Y, Yoon YW, Na HS, Kim SH, Chung JM. Behavioural signs of ongoing pain and cold allodynia in a rat model of neuropathic pain. *Pain* 1994; 59:369–376.

Dai Q, Zhou D, Xu L, Song X. Curcumin alleviates rheumatoid arthritis-induced inflammation and synovial hyperplasia by targeting mTOR pathway in rats. *Drug Des Devel Ther* 2018; 12:4095–4105.

Hargreaves K, Dubner R, Brown F, Flores C, Joris J. A new and sensitive method for measuring thermal nociception in cutaneous hyperalgesia. *Pain* 1988; 32:77–88:32.

Jasemian Y, Svendsen P, Deleuran B, Dagnees-Hansen F. Refinement of the collagen induced arthritis model in rats by infrared thermography. *British journal of medicine and Medical Research* 2011; 1(4):469–477.

Kim HY, Kim WU, Cho ML, Lee SK, Youn J, Kim SI, et al. Enhanced T cell proliferative response to type II collagen and synthetic peptide CII (255-274) in patients with rheumatoid arthritis. *Arthritis Rheum* 1999; 42:2085–2093.

Kim WU, Cho ML, Jung YO, Min SY, Park SW, Min DJ, et al. Type II collagen autoimmunity in rheumatoid arthritis. *Am J Med Sci* 2004; 327:202–211.

Koizumi F, Matsuno H, Wakakaki K Synovitis in rheumatoid arhtrtis: scoring of characteristic histopathological features. *Pathology International* 1999; 49:298–304.

Krenn V, Morawietz L, Burmester GR, Kinne RW, Mueller-Ladner U, Muller B, Haupl T. Synovitis score: discrimination between chronic low-grade and high-grade synovitis. *Histopathology* 2006; 49:358–364.

Lim MA, Louie B, Ford D, Heath K, Cha P, Lacroix JB, Lum PY, Robertson TL, Schaevitz L. Development of disease arthritis index, a novel metric to measure disease parameters in a Rat model of Rheumatoid arthritis. *Front Pharmacol* 2017; 8:818.

Mikecz K, Glant TT, Buzás E, Poole AR. Proteoglycan-induced polyarthritis and spondylitis adoptively transferred to naive (nonimmunized) BALB/c mice. *Arthritis Rheum* 1990; 33:866–876.

Nosratiz, BM, Rodriguez CR, Saatchii K, Hafeli UO Refinement and Validation of infrared thermal imaging (IRT): A non-invasive technique to measure disease activity in a mouse model of rheumatoid arthritis. *Arthritis Research & Therapy* 2020; 22:281.

Pearson CM. Development of arthritis, periarthritis and periostitis in rats given adjuvants. *Proceedings of the Society for Experimental Biology and Medicine* 1956; 91(1):95–101.

Sanchez BM, Lesch M, Brammer D, Bove SE, Thiel M, Kilgore KS. Use of a portable thermal imaging unit as a rapid, quantitative method of evaluating inflammation and experimental arthritis. *J Pharmacol Toxicol Methods* 2008; 57(3):169–175.

Snekhalatha U, Anburajan M, Venkatraman B, Menaga M, Raj B. Evaluation of Complete Freund Adjuvant Induced Arthrtitis model in Wistar rats by Thermography, A comparison with histopathology. *Zeitschrift fur Rheumatologie* 2012; 72(4):375–382.

Tal M, Bennett GJ. Extra-territorial pain in rats with a peripheral mononeuropathy: mechano-hyperalgesia and mechano-allodynia in the territory of an uninjured nerve. *Pain* 1994; 57:375–382.

Tuncel J, Haag S, Hoffmann MH, Yau ACY, Hultqvist M, Olofsson P, et al. Animal models of rheumatoid arthritis (I): Pristane-induced arthritis in the rat. *PLoS ONE* 2016; 11(5):e0155936. doi:10.1371/journal.pone.0155936

Wooley PH, Siegner SW, Whalen JD, Karvonen RL, Fernández-Madrid F. Dependence of proteoglycan induced arthritis in BALB/c mice on the development of autoantibodies to high density proteoglycans. *Ann Rheum Dis* 1992; 51:983–991.

Xia C, Li J, Yue Y, Chen L, Dai J. Biophoton imaging evaluation of the process of rheumatoid arthritis in rats. *Natural Science*, 2021; 13(10):451–456.

Yu Y, Xiong Z, Lv Y, Qian Y, Jiang S, Tian Y. In vivo evaluation of early disease progression by X-ray phase-contrast imaging in the adjuvant-induced arthritic rat *Skeletal Radiol* 2006; 35:156–164.

5 Thermal Imaging for Inflammatory Arthritis Evaluation

Thermal imaging is a non-contact, non-invasive, diagnostic imaging procedure based on skin temperature measurement. It is a functional imaging method for analyzing physiological functions related to body temperature. It has been used as a pre-screening tool in various clinical applications of rheumatology, such as diagnosis of patellofemoral arthralgia in knee joint (Devereaux et al., 1986), monitoring of skin temperature elevations in scleroderma (Maeda et al., 1998), rheumatoid arthritis (RA), (Devereaux et al., 1985), knee osteoarthritis (Warashina et al., 2002), and juvenile idiopathic arthritis (Spalding et al., 2008).

Thermal imaging contributes the fact about the regional inflammatory process which affects the superficial dermal blood circulation. The exchanges of heat radiation from tissue or organs to the skin surface are being substantiated during the pathological conditions. Due to the occurrence of the elevated or decreased superficial temperature during the disease progression, a decline in inflammation is observed. These profound temperature variations can be observed using infrared thermal cameras and effectively assessed after the pharmacological therapeutical process. In rheumatoid arthritis, thermal imaging is used to assess the elevated temperature pertinent to inflammation or reduction in temperature due to blockage in blood vessels.

The advent of emerging technologies in artificial intelligence plays a predominant role in the detection of subtle changes in thermal imaging applications. Hence, the computer vision comprises image segmentation and analysis techniques as well as feature extraction process followed by classification using machine learning classifiers. This chapter is focused on the application of various image segmentation techniques in thermal images of arthritis and feature point detector based on ORB and categorizing the arthritis and normal using various machine learning techniques.

The study aims are as follows: (i) to study the health assessment scores and temperature indices of rheumatoid arthritis patients, (ii) to apply various image segmentation techniques in hand thermal imaging and identify the abnormal region of interest (ROI), and (iii) to extract the hand-crafted features based on advanced feature extraction methods and classify the RA and normal using various machine learning classifiers.

DOI: 10.1201/9781003245780-5

5.1 PATIENTS AND METHODS

The free arthritis screening camp was conducted in the month of October 2021 for the population living in and around Chengalpattu district, at SRM Medical College Hospital and Research Centre, Kattankulathur, Chennai, Tamil Nadu, India. A detailed questionnaire which contains information about the demographic variables such as age, disease duration, biochemical variables, anthropometric information such as height, weight, and body mass index (BMI), and a full health assessment such as health assessment questionnaire (HAQ) score and disease activity index score or disease activity score (DAS) was filled out for each participant. The consent form signed by the patient was collected for all the participants involved in the study. The Institutional Ethical Committee, SRM Medical College Hospital and Research Centre has approved the study.

According to the Indian Rheumatology Association (IRA) consensus report (2008), subjects with inflammatory arthritis in more than four joints, elevated erythrocyte sedimentation rate (ESR)/C-reactive protein (CRP), positive IgM RF, radiographic changes of juxta-articular osteopenia, erosions, and joint space narrowing were confirmed as RA (Mishra et al., 2008). In this study, the confirmed RA participants and age-matched controls were included. The participants who had undergone, physiotherapy, hand fractures, and fever were excluded from the study. The age group of the participants ranged from 30 to 75 years. About 80 subjects participated in the study, and 20 subjects were excluded because of other co-morbidities such as diabetes and CVD. The details of the study population were as follows:

- Total subjects: 60
- RA (n = 30, M/F ratio = 1:3, age = 47.4 ± 10.4 years, disease duration = 4.5 ± 2.5 years)
- Age- and sex-matched normal (n = 30, M/F ratio = 1:3, age = 46.5 ± 10.5 years)

The thermal image was captured using a portable thermal camera (ThermaCAM A305Sc, FLIR, Sweden) in both dorsal and ventral views. The camera employs 320 × 420 thermal resolution with an uncooled microbolometer detection system with an accuracy of 2°C. The participants were advised to sit with their hands exposed for duration of 10 minutes in a temperature-controlled room at 20°C with a humidity of 50%. During the thermal image acquisition process, the distance between the thermal camera and participant's hand was kept at 1 m. To avoid any artifacts, the images were acquired in a dark room with the background properly covered with black color cloth. FLIR software Version 1.2 is used for analyzing the temperature changes in the hand thermal images. Further, the images were analyzed for segmentation and classification using MATLAB 2020 software and Python programming.

5.2 DISEASE ACTIVITY ASSESSMENT

With the arrival of biologicals and the concept of "treat to target," there was an increased need for tools of quantified assessment of disease activity in RA patients (Anderson et al., 2011). Clinical features such as the number of tender and swollen joints were combined with blood tests and general health assessments to generate various DASs.

HAQ is an outcome measure introduced in 1982 (Fries et al., 1982) to quantitatively evaluate the functional ability and the daily living activities of RA patients. The HAQ consists of 20 questions, which cover eight categories of daily life activities. Questions can be answered with "possible without any difficulty (0 point)," "possible with some difficulty (1 point)," "possible with much difficulty (2 points)," and "unable to do (3 points)." Use of any aids increases the score by 1 point. The worst score of each category is accepted as a representative value for this activity. The total HAQ score is equal to the mean value of all sub-scores. Interpretation of mean HAQ scores should consider the following.

A score of 0.5 means that the scoring sum of the eight domains is 4 points, which can be the result of the following combinations:

- 4 times 1 (= possible with some difficulty),
- 2 times 1 and 1 times 2 (= possible with much difficulty),
- 2 times 2 (= possible with much difficulty), or
- 1 time 1 and 1 times 3 (= impossible to do). 3 points can also be the result of using an aid plus scoring 2 points.

The fact that a task is impossible to perform cannot be qualified as a low level of disablement and consequently, the cut-off for slight HAQ disability should be set to 0.25. At a level of 0.24, the HAQ instrument shows a floor effect and cannot reliably differentiate slight disabilities. Normative values for the HAQ-DI are available from Finland (Krishnan et al., 2004). A mean HAQ of 0.25 (95% confidence interval: 0.22–0.28) was reported. In addition to the well-known association between HAQ and disease activity, a strong correlation is established between the level of physical activity and HAQ. There are three evaluated versions of the HAQ score. Ten different disease activity measures included HAQ data into their database for creating a DAS (Anderson et al., 2011).

(a) **Disease Activity Score:** The DAS was calculated for confirmed RA patients based on their clinical and laboratory data. The DAS28 score includes a 28 swollen joint count, a 28 tender joint count, ESR, and a patient global assessment of disease activity or general health on a visual analog scale (VAS). DAS28 may also be calculated using the CRP level instead of Westergren ESR. There are formulas to calculate the score of all four DAS28 versions, which can also be obtained using the online calculator available on the Internet (http://www.das-score.nl). The DAS28 score was categorized as follows: (i) DAS28 > 5.1 represents person with high disease activity, (ii) 3.2 < DAS28 < 5.1 indicates moderate disease activity, (iii) 2.6 < DAS28 < 3.2 depicts low disease activity, and (iv) DAS28 < 2.6 indicates remission due to absent inflammatory disease activity.

(b) **Clinical Disease Activity Index**: Clinical disease activity index (CDAI) is a clinical composite score based on four parameters: a 28 swollen joint count, 28 tender joint count, patient global assessment of disease activity on a 10-cm VAS, and provider global assessment of disease activity on a 10-cm VAS. The CDAI is calculated by adding the four items together. The CDAI scores are grouped as follows (Ghosh et al., 2011):

A score in the range 22.1–76 indicates high activity,
10.1–22 represents moderate activity,
2.9–10 shows the low activity, and
0–2.8 denotes the remission.

The simplified disease activity index (SDAI) is analogous to the CDAI but extended by the laboratory measurement of CRP level in mg/dL. Score calculation is CDAI score plus CRP value, of which a maximum value of 10 mg/dL is often defined. Thus, the upper limit of SDAI is 86 points. The SDAI represents disease severity as follows (Ghosh et al., 2011):

1. 26.1–86 indicate high activity,
2. 11.1–26 represents moderate activity,
3. 3.4–11 shows low activity, and
4. 0–3.3 represents remission.

5.3 TEMPERATURE INDICES

The temperature indices were evaluated quantitatively to measure the temperature parameters of RA patients (Collins et al. 1974). The parameters measured are as follows: skin surface temperature, thermographic index, and heat distribution index (HDI).

A square ROI of 10 × 10 mm size was fixed in the hand thermal images of metacarpophalangeal (MCP), proximal interphalangeal (PIP), and distal interphalangeal (DIP) joints. Skin surface temperatures such as maximum, minimum, and average temperatures were displayed for both the dorsal view and ventral view of the hand region in the FLIR software.

The thermographic index (TI) measures the quantitative thermal changes in ROI, particularly, displaying the mean joint temperature in each area of the temperature band using the following equation:

$$TI = \sum \frac{\Delta T \, X \, a}{A} \qquad (5.1)$$

where ΔT represents the difference between the measured isothermal temperature and a constant, 26°C, "a" indicates the isothermal area (cm^2), and "A" represents the total thermogram area (cm^2) (Collins et al. 1974). The TI value is categorized as follows: (i) 2 – normal, (ii) 3–4 – osteoarthritis, and (iii) >4 – rheumatoid arthritis (Ring, 1980).

5.3.1 HEAT DISTRIBUTION INDEX

HDI was developed in the early 1980s at the Rheumatology Department, Addenbrooke's Hospital, Cambridge, UK, and represents the distribution of temperature across the joints (Salisbury et al., 1983). The HDI was determined in thermograms of the knee, the ankle, and elbow recorded in a lateral view from healthy controls and patients with inflammatory arthritis. The authors defined the normal thermal pattern over a joint as one in which there is a negative temperature gradient from the center to the periphery of the joint without localized areas of increased temperature. In general, this gradient is of the order of 2–3°C. The relative frequency distribution (RFD) of temperatures at the investigated joint was calculated as the number of pixels in the ROI which occur at each temperature level.

$$\text{RFD} = \frac{\text{Probabilty of occurrence of particular pixels}}{\text{Total number of pixels}} \tag{5.2}$$

The HDI represents the width of the RFD equal to ± 1 standard deviation (SD) from the mean frequency; thus, HDI indicates the variation, but not the mean temperatures of an ROI.

The authors compared the HDI and thermographic index with clinical assessment in arthritic knee joints. They observed an approximately linear relationship between HDI and clinical assessment. The authors also found that a good correlation exists between the TI and clinical assessment.

Devereaux et al., (1985a, 1985b) combined the HDIs from a number of medium-sized joints such as wrist, elbows, knee, and ankle and correlated thermography with other assessments of rheumatoid activity such as VAS for the assessment of morning stiffness and pain, grip strength, the Ritchie articular index, hemoglobin, ESR, and CRP. The study results confirmed the expected significant correlation between thermography and other clinical parameters The authors recommended the HDI as a method for the assessment of response to therapy in patients with rheumatoid arthritis.

Spalding et al. (2008) conducted a proof-of-concept study to determine whether two of the cardinal signs of disease activity in arthritis (swelling and warmth) can be reliably quantified using existing three-dimensional (3D) and thermal digital imaging devices. For detection of surface changes due to joint swelling, the authors used a laser line triangulation scanner to acquire 3D images from the wrist and MCP region of arthritic subjects and controls. They used Rapidform software for constructing the 3D models from the scanned images. ROIs were fixed in the 2nd–5th MCP and wrist region and the authors calculated the volume within each ROI. Then, the surface distribution index (SDI) was generated using the in-built function named "shell–surface deviation" of the Rapidform software. SDI data reflect the standard deviation from the mean distances from surface points to the bottom plane. Good reliability of volume measures was reported in repeated measurements of a small sample of clinical cases. Based on these results, volume changes greater than 1.1 mL in the wrist and 0.5 mL in the MCPs between imaging sessions would be considered significant. It was observed in a clay-generated mannequin hand that a significant increase from baseline volume was detectable with the addition of as little as 0.2 mL of clay (0.8% above baseline volume) to the MCP ROI and 0.6 mL (1.5% above baseline volume) to the wrist ROI. The SDI was able to discriminate between focal and diffuse swelling when 1.6 mL of clay was added to the wrist ROI ($p = 0.02$) and 0.6 mL to the MCP ROI.

For thermal evaluation, the authors applied the HDI defined as twice the standard deviation of the mean of all temperatures within the required ROI. The HDI showed little variation in measurements repeated on the same day of non-arthritic hands. The variation was much higher in measurements on different days, but no control wrist or MCP HDI exceeded 1.3°C. The authors suggested that the HDI above 1.3°C is considered indicative of the presence of the arthritis and confirmed this discriminative threshold by comparing wrists and MCPs of healthy controls with the corresponding joints of arthritis patients.

The HDI can lead to erroneous results if small joints present with homogeneously distributed high temperatures. In this case, the standard deviation of the mean temperature will be small and the HDI may miss an inflammatory condition despite an absolute temperature suspicious for arthritis.

5.4   THERMAL IMAGE ACQUISITION PROTOCOL

Before the image acquisition process, acclimation time to a room temperature of 20 ± 0.5°C for 15 minutes was observed for each participant. Preparation of patients: for capturing thermal images of hands or feet, the hands and forearms, feet, and lower legs must be undressed (undressing must include jewelry such as finger rings). For elbow and shoulder joints, the upper extremities and the upper body must be undressed (inner fit in females may remain). For ankle joints, the same undressing rule applies as in feet imaging; knees need undressing from the foot to the upper thigh; and the hip joint and the gluteal region requires the removal of underpants.

During the acclimation period, the persons being imaged should sit or stand for images of the upper extremities or the knee, for images of the hip region the individual should stand, and for foot and ankle images, they should sit with their feet placed on a support to avoid contact with the floor.

During acclimation, the person being imaged should move as little as possible and have their hands hanging freely on side of the body, not touching parts of their body or any other surfaces.

For detection of inflammatory arthritis, images of both hands and both feet should be captured in a dorsal view. This requires placing the hands on a support (table) with contact at the palms. The camera must be positioned perpendicularly above the dorsal surface of the hands. This requires special camera mounting facilities available with a single pillar photographic stand.

Such a camera support is also helpful in capturing a thermal image of both dorsal feet. The individual should sit on a chair, legs slightly abducted, feet placed parallel on the floor, and the big toes pointing forward and knees flexed to 80°. The camera is placed at knee height, looking perpendicularly at the feet.

Both knees and ankles are imaged in an anterior close-up view. For knees, the person stands with legs slightly abducted, knees in full extension, feet parallel, and the big toes pointing toward the camera. The camera looks at knee height perpendicularly to the anterior knee surface, with the upper edge of the field of view 3.5 cm above the rim of the patella and the lower edge 2.5cm below the tip of the fibula.

The same body position is recommended for thermal imaging of ankles, but the person should be positioned on a low stage or two-level staircase, to allow a perpendicular camera view, since the lowest possible level is at a 20–30 cm distance to the floor.

Camera views and related conditions for image capturing are clearly described in the book *The Thermal Human Body* (Ammer & Ring, 2019). This chapter also includes information on how and where ROIs should be defined in thermal images.

Due to typical distribution patterns of inflamed hand joints in rheumatoid arthritis patients, only ROIs over the wrists, the MCP joints, and the PIP joints have diagnostic relevance.

For evaluation, the maximum, minimum, mean temperatures, and the standard deviation of the mean are extracted from the ROIs. These temperatures are then compared to the temperature in corresponding ROIs on the contralateral side. Differences in mean temperature greater than 0.5° are suspicious of abnormality and a difference of 1°C is considered abnormal.

In case of bilaterally high temperatures, the thermal index (TI) helps to detect temperatures that are associated with inflammation. Since the TI is generated by subtracting a mean skin temperature from the mean temperature measured over joints, a TI > 4°C indicates abnormality even in bilateral involvement. The HDI can help to better define the absolute temperature at the edge of abnormality, because a high HDI in an ROI with a mean temperature exceeding the contralateral ROI by 0.5°C may already indicate evolving inflammation.

All steps of image analysis must be performed manually, by trained experts, and is time-consuming and difficult. Computer-based systems are available for more than two decades to assist in the analysis of thermal images. However, all these approaches are still experimental in nature and consensus on standard software is not yet achieved.

For a rheumatologist, software assisting the analysis of thermal imaging must meet the following requirements:

A. **Definition of Regions of Interest**:
  1. Automatic definition of over joints in standard views of body regions with an option to correct shape, size, and location of the ROI.

2. Option to manually define ROIs of any possible shape, size, and location in a recorded thermal image.
3. Option of defining cross-sectional lines of any possible length, orientation, and location.
4. Option to manually select ROIs from the automatically offered set.
5. Option to define relative temperatures between any defined ROIs.

B. **Display of Results**:

1. Automatic display of temperatures (mean ± standard deviation, minimum, maximum; TI, HDI) within each ROI and of cross sections.
2. Automatic display of the size of each ROI and cross sections in terms of numbers of pixels.
3. Automatic display of relative temperatures defined as the absolute difference of mean temperature and standard deviation of corresponding ROIs.
4. Since many problems of analyzing thermal images arise from neglecting standards during image capture, resulting in poor images, a software-based guide would improve the quality of acquired thermal images.

Such systems could control the following prior to image capture:

1. The background temperature within the thermal image, which should ideally be the same as the temperature in the examination room (according to that, the images a and b of Figure 5.4 were recorded at an ambient temperature between 32 and 33°C).
2. Set the upper limit of the temperature window to 37°C. This will easily detect outliers since, at a room temperature of 20°C, skin temperatures do not exceed 36°C and colors in the red-to-white portion of the color bar represent most likely false readings.
3. Provide masks representing the outline of an ideally positioned body region. Prior to image capture, the mask should be matched with the actual object in a software-assisted way.
4. Images not fulfilling the requirements for standard views should be labeled as poor and archiving should be rejected. Such a quality assurance process requires the same image processing techniques as necessary for image segmentation and image registration.

Figure 5.1 shows the thermal image acquired in the dorsal view for an RA patient with proper background environment. The ROI is fixed as constant and used square tool to locate the MCP and PIP joints of all the fingers. Hence, in total, ten regions are fixed as ROIs and the average temperature in each ROI is noted down. Similarly, the same ROI is fixed in hand image of a normal subject as depicted in Figure 5.2. Figure 5.3a indicates the thermal image of an RA patient in which when square ROI is kept on the 3rd MCP joint, the minimum, maximum, and average temperatures are displayed. Similarly, Figure 5.3b represents the thermal

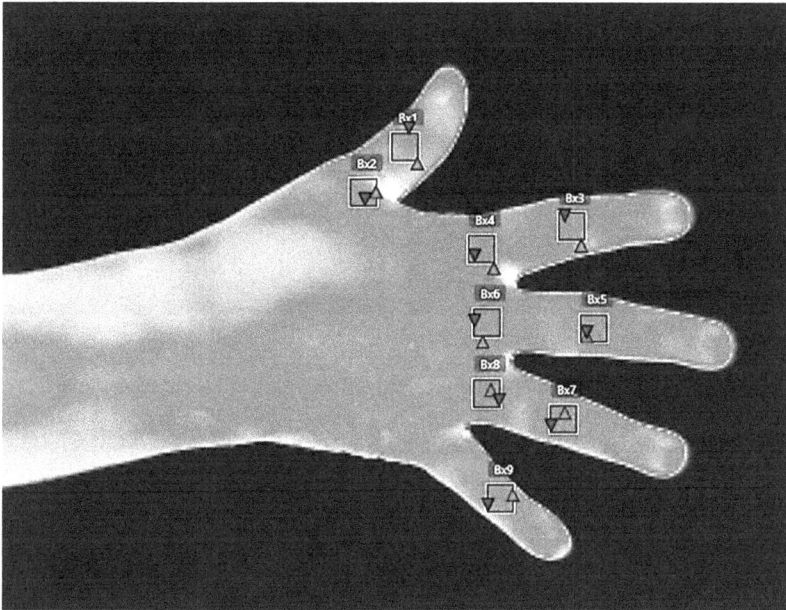

FIGURE 5.1   Temperature measurement performed in hand joints of thermogram of RA patients.

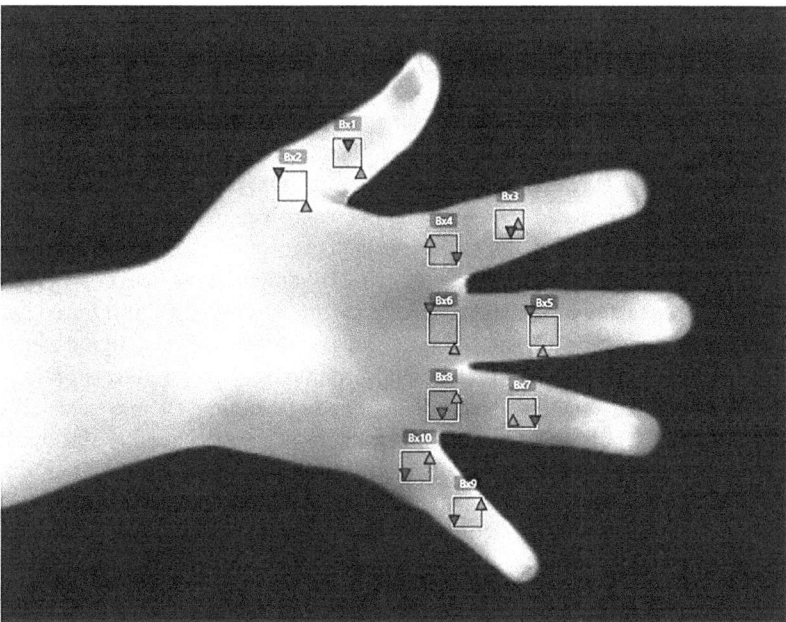

FIGURE 5.2   Normal thermal image in which a constant ROI is placed in MCP and PIP joints of all the fingers.

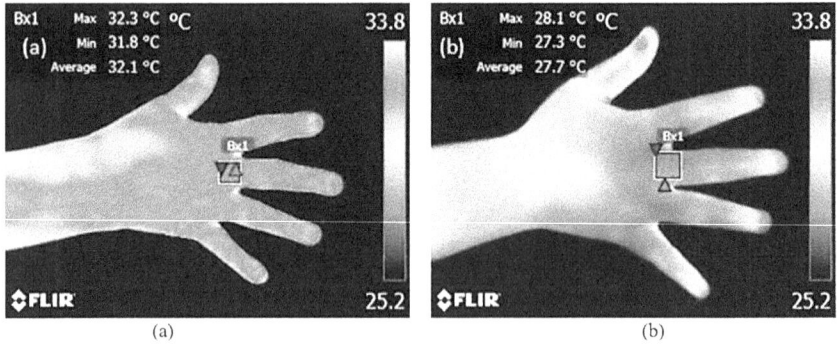

(a)   (b)

FIGURE 5.3 (a) Thermal image of an RA patient in which ROI is fixed in the MCP joint as the square box. (b) Normal thermal image in which square ROI is placed in the MCP joint of the third finger.

image of normal subjects depicting the average temperature. It was observed that the difference in temperature between RA and normal in the 3rd MCP joint is found to be 1.15°C and the percentage difference was obtained as 3.6% between them using the following formula:

$$\% \text{ difference} = \frac{\text{RA} - \text{Normal}}{\text{RA}} \times 100 \qquad (5.3)$$

5.5 THERMAL IMAGE SEGMENTATION

Image segmentation plays an important role in image analysis and may be described as process wherein background objects are separated from the desired section of an image (Preetha et al., 2012). Thermal images depend on the detection of radiation in the infra-red region of the spectrum (Nandhakumar and Aggarwal, 1988). According to black body radiation, each object emanates thermal radiation since its temperature is above absolute zero. For temperatures between ~100 and 1000 K, the wavelength of the thermal rays is in the range of the infrared spectrum. The advantages of infrared imaging cameras over traditional digital cameras are that infrared cameras work also in environments with low levels of visible light. These include security, surveillance, and packaging (Wang et al., 2010, Chen et al., 2003). Typically, a thermal camera is used to acquire the thermal image of the desired body part or ROI such as a subject's foot, hand, or individual digits. Then, the images are captured and stored using the dedicated software after which the images are pre-processed before being subjected to some form of image segmentation. There are several image segmentation methods that can potentially be used for analyzing the thermal images. These include threshold-based segmentation, edge detection method, k-means clustering, and watershed segmentation (Nandagopan and Haripriya, 2016).

5.5.1 THRESHOLD METHOD

When there is a large difference between the background and the gray set of the target image, the simple threshold method can be applied (Jadin 2012). The gray level histogram needs to be computed, after which the in-class variance value of the two-color gradients (the foreground and background) need to be calculated. The optimum threshold value is the in-class value that is the smallest. This threshold value is used to convert to a black and white, dual-color system. Using this, the largest object can be determined using fields that are formed independently, by interconnecting pixels (Kacmaz et al., 2017).

$g(x,y)$ can be used to represent a threshold image and the function $f(x,y)$ can be used to represent the thermal image (Suma et al., 2016). The relation is as follows:

$$g(x, y) = 1, f(x, y) > T \tag{5.4}$$

$$0, f(x, y) < T$$

Threshold-based methods can be used to segment the hot spot regions from thermal images. Segregating an image into sections based on shared features of the pixels in the image makes it easier to apply image processing techniques to said regions (Shaikh et al., 2016).

Finding the ROI of the thermal image using segmentation techniques is a challenging task due to the captured image having a smooth variance of gray shades, which means there are no sharp changes in the images. The utilization of global thresholding algorithms can mitigate this issue. The global thresholding method works from the assumption that the image has a histogram with two modes. As a result, the binary image is produced after thresholding (Shaikh et al., 2016). The images that are global thresholding-based provide 8-bit and 16-bit images in a binary format. Different threshold-based segmentation techniques yield varying results when utilized on thermal images. Manual thresholding was applied for the segmentation of thermal images of RA patients. Initially, the input thermal image was converted into a grayscale image. The threshold value of T = 95 was applied on the grayscale image based on trial and error method manually. In the automated method of thresholding process, the sum of the pixel values was computed for the grayscale image. Then, the sum of the pixel value was divided by the size of the grayscale image to yield the average pixel value. Later on, the automated threshold value was computed based on the condition that if the pixel value of the grayscale image is greater than the average pixel value, the segmented binary image is obtained.

The segmented image contains the foreground object as the hot spot region whose intensities are black and other regions are displayed as white regions. When compared to the manual method of thresholding, the automated thresholding method provided better results in segmentation in which small intensity

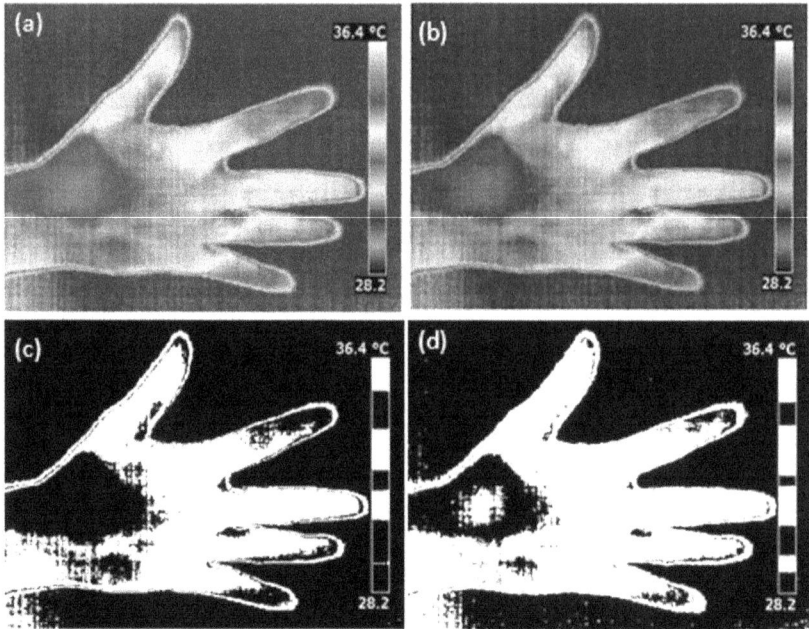

FIGURE 5.4 Threshold-based segmentation of hot spot region in RA patient thermal imaging: (a) input thermal image, (b) grayscale image, (c) manual thresholded image, and (d) automated thresholded image.

variations are displayed as indicated in Figure 5.4. Here, Figure 5.4 represents the threshold-based segmentation applied to the hand region. Both automated and manual methods of thresholding are applied on the thermal images. Figure 5.4a shows the original thermal image of the RA patient, Figure 5.4b represents the thermal image converted into the grayscale image, Figure 5.4c indicates the manual thresholding with the threshold value set at T = 95, and Figure 5.4d depicts the automated thresholding applied to the grayscale image.

5.5.2 Edge Detection

The detection of abrupt changes in the brightness of an image which are called discontinuities is called edge detection. Edges or boundaries refer to points where image brightness varies sharply. Edge detection has applications in pattern recognition, machine vision, and image processing. Commonly used edge detection methods include Sobel, Canny, and Prewitt. Smoothing and differentiation of the image are usually required for edge detection methods (Duarte et al., 2014; Kadir and Brady, 2001, Sergyan, 2012). Edge detection can be used to detect rheumatoid arthritis from the thermal images of small animal models. Several edge detection models can be used for this purpose such as the Robert edge detector and Canny edge detector. In the case of the Robert edge detection method, a 2D spatial gradient measurement is performed on the thermal image, after which the Robert operator is applied. The gradient magnitude is found and the original pixel values are replaced by the magnitude. The image is convolved and the edge detected image is formed (Snekhalatha et al., 2011).

The convolution operation is performed by placing the kernel over the input image, multiplying the kernel value of 3 × 3 with the input pixel values followed by summation. Hence, the operation is carried out for the entire image by moving the kernel over the input image. To prevent the loss of information, zeros are padded around the input image both vertically and horizontally on all sides. Then, the convolution operation is performed between the input image pixels and the kernel. The resultant pixel values are considered as the convoluted image. Figure 5.5a illustrates the sample convolution operation performed between the input image and the 3 × 3 kernel. Figure 5.5b shows the entire input image convoluted with the 3 × 3 kernel and the resultant convoluted image is displayed.

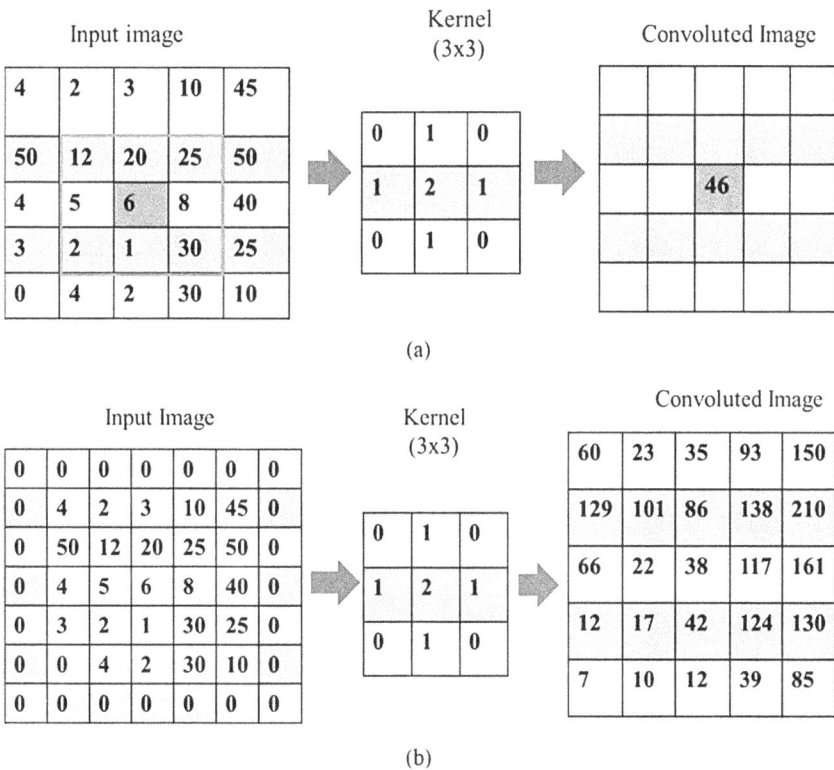

(a)

(b)

FIGURE 5.5 (a) An example of convolution operation performed between input image and kernel. (b) Convoluted image obtained after performing image convolution operation of the input image with a 3 × 3 kernel.

For the Canny edge detection method, the original image is smoothened, and then the intensity gradient of the image is found. The image then undergoes non-maxima suppression, thresholding, and edge tracking. This is how the edge detected image is formed using this technique (Gonzalez and Woods, 2007). It was observed that edge detection techniques fail to detect proper abnormal ROIs due to the discontinuities present in the detected edges of the image. Among all the edge detection techniques applied, the Canny edge detector and the log edge detector detect many unwanted edges present in the image. Hence, these simple edge detection techniques may not be suitable for segmenting the hotspot ROIs in abnormal arthritis images. It is advisable to adopt advanced segmentation techniques which could be capable of detecting the abnormal ROIs. Figure 5.6 illustrates the different types of edge detection techniques applied to thermal imaging of RA patients. Figure 5.6a shows the input hand thermal image in the dorsal view of an RA patient. Figure 5.6b represents the input thermal image converted into grayscale image, Figure 5.6c indicates the grayscale image resized to a standard size of 256×256, Figure 5.6d depicts the image obtained after the Robert edge detection operator was applied on the resized grayscale image, Figure 5.6e displays the image attained after the application of the Prewitt edge detection operator, Figure 5.6f portrays the output image obtained after Sobel edge detection operator was applied, Figure 5.6g represents the Canny edge detection output, Figure 5.6h indicates the resultant image obtained due to the application of the log edge detection operator, and Figure 5.6i shows the image attained due to the zero cross edge detection operator.

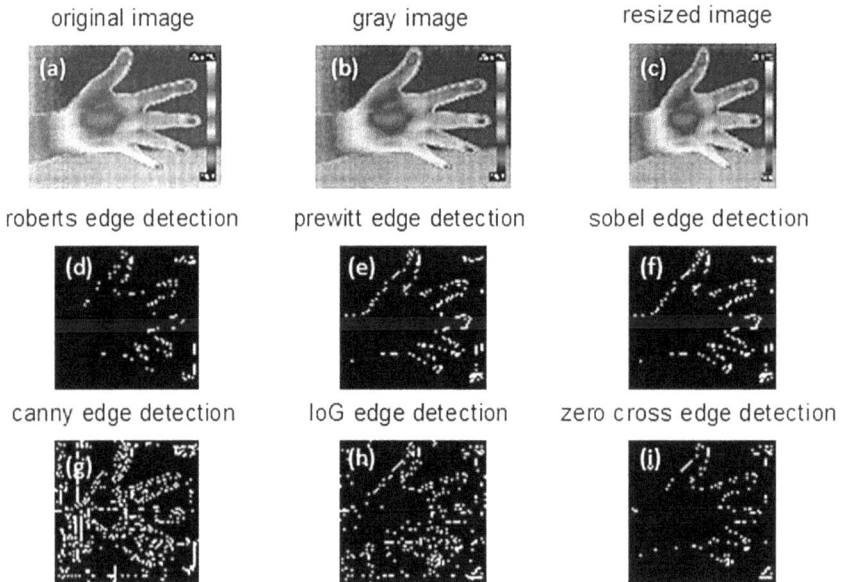

FIGURE 5.6 Different types of edge detection operators applied to thermal imaging. (a) input thermal image, (b) gray scale image, (c) resized image, (d) Robert edge detected image, (e) Prewitt edge detected image, (f) sobel edge detected image, (g) canny edge detected image, (h) log edge detected image, and (i) zero cross edge detected image.

5.5.3 k-Means Clustering

k-means clustering is an unsupervised iterative and heuristic partitioning algorithm used to divide a large group of data points into smaller groups. This makes the technique valuable in image processing, pattern recognition, etc. (Nameirakpam et al., 2015). Clustering is the division of similar groups of data. k-means is a commonly used form of clustering that relies on counting the number of clusters present within the given data and by specifying the cluster seeds, which are starting cluster centroids. They are either randomly chosen or the data scientist specifies it, depending on their knowledge of the given data. The data points are then assigned to a seed based on proximity. The centroid is then calculated and the prior two steps are repeated until the clusters are no longer adjustable. If the value of k is not known, then the cluster centroid will not lie inside the clusters if the value of k is too small (Likas et al., 2003). k-means clustering has several applications in medical imaging, such as object extraction where k-means clustering was used for the segmentation of hot spot regions in the thermal image of arthritis (Snekhalatha et al., 2015, 2017). The k-means clustering algorithm is applied to the thermal image of RA patients. In this proposed work, three clusters are initialized with $k = 3$, where cluster 1 indicates the hot spot abnormal region, cluster 2 represents all the other regions including the background, and cluster 3 shows the background region. After calculating the cluster center from each cluster, the distance between the input image and cluster center is computed, based on the minimum distance criteria, and the pixel values in the input image get segregated into different cluster regions. Figure 5.7 indicates the k-means clustering algorithm applied on the thermal imaging of RA patients. Figure 5.7a represents the original thermal image of RA patients, Figure 5.7b shows the cluster 1 depicting the hotspot region,

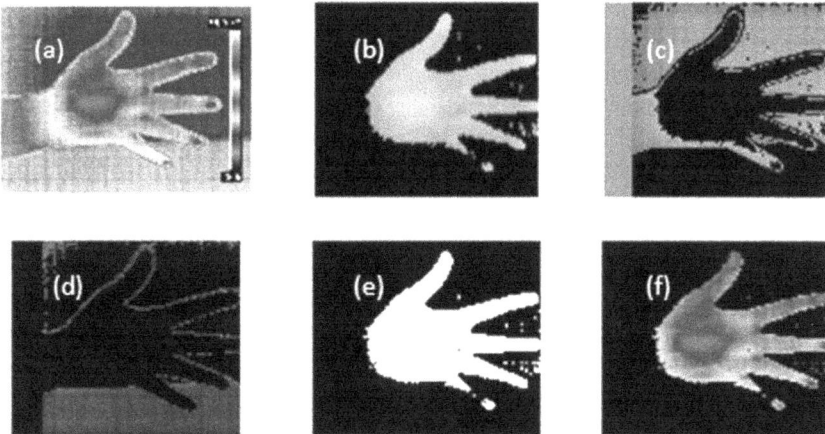

FIGURE 5.7 Various clusters and segmented output obtained by application of k-means clustering algorithm in thermal imaging. (a) input thermal image of RA patients, (b) cluster 1 depicting the hotspot region, (c) cluster 2 portrays the regions other than hotspot areas, (d) cluster 3 depicting the background region, (e) binary thresholded image, and (f) segmented hot spot region,

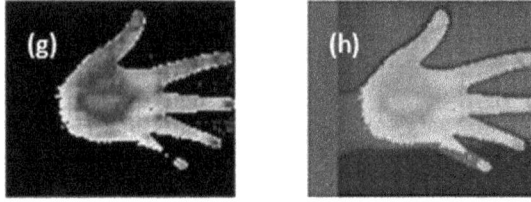

FIGURE 5.7 (*Continued*) (g) grayscale conversion of the segmented image, and (h) segmented image superimposed with the original image.

Figure 5.7c displays the cluster 2 portrays the regions other than hotspot areas, Figure 5.7d represents the cluster 3 depicting the background region, Figure 5.7e shows the binary thresholded image, Figure 5.7f depicts the segmented hot spot region, Figure 5.7g indicates the grayscale conversion of the segmented image, and Figure 5.7h indicates the segmented image superimposed with the original image.

5.5.4 Watershed Segmentation

Traditionally, watersheds are ridges that help divide areas drained by different river systems. Image segmentation is the separation of objects from each other and the background. User-defined markers are used to choose local minima for the image. The pixel values are treated like local topography (Ramirez-Rozo et al., 2012). The image must be visualized as 3D with watershed segmentation. These are the two spatial coordinates versus gray levels (Chaudhary and Chaturvedi, 2017).

Watershed segmentation can be very useful in tumor detection as tumors are highly dense due to them containing proteinaceous fluid. As watershed segmentation can segregate an image depending on its intensity portions, it acts as a very good way to help differentiate between normal and tumorous tissues in thermal images. In the watershed segmentation method, the input thermal image is converted into a grayscale image, then the Sobel operator is used to extract the horizontal and vertical edge of the grayscale image. Then, the gradient magnitude is obtained by taking the square root of squaring the horizontal and vertical edges together. Then, the watershed is applied to the gradient magnitude image. The image appears with different gray shades as indicated in Figure 5.8. Then, a morphological operator such as the image opening operator is applied on the grayscale image along with "disk" as the structuring element. Then, the resultant image obtained is a convolved image. The erosion operation is applied on the convolved image and the image reconstruction was performed by combining the erosion image and grayscale image. Hence, the operation resultant from that is termed as image opening by reconstruction. Then, the image closing operation was performed by convolving the image opening operator with the structuring element. The dilation operation is applied to the combination of the reconstructed image with the structuring element. The reconstructed image is obtained by taking the complement of the dilated image. Thus, we obtain the image opening and image closing operation by reconstruction. Then, a regional maximum is applied on the image opening–closing by

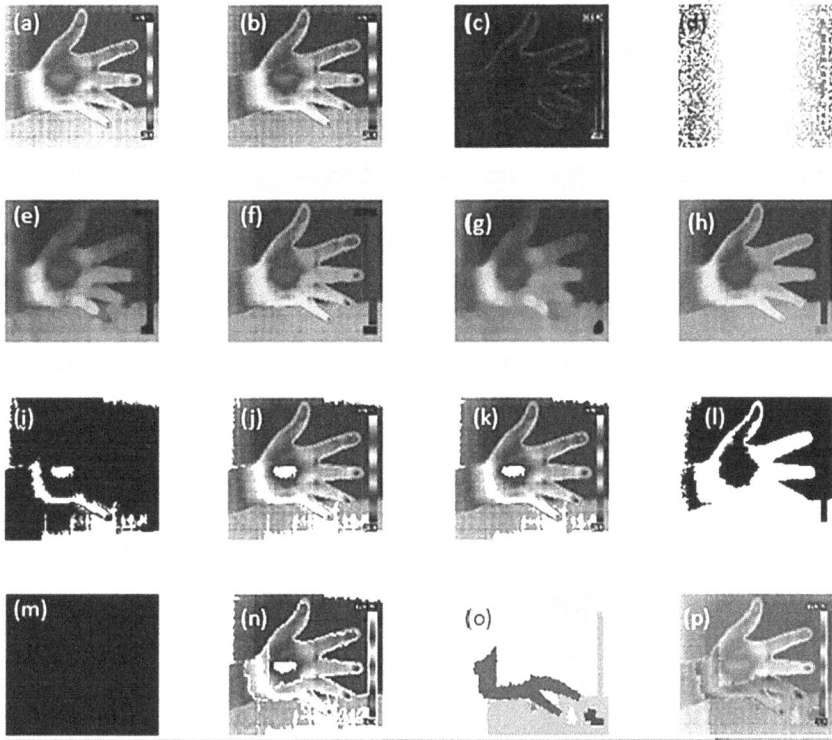

FIGURE 5.8 Watershed segmentation techniques applied in thermal imaging of RA patients. (a) Input thermal image of RA patients, (b) grayscale image, (c) gradient magnitude image, (d) watershed transform of gradient magnitude image, (e) image opening operation applied on the transformed image, (f) image opening operator followed by the application of image reconstruction operation. (g) combination of image opening and closing operation applied on the image, (h) image opening–closing operation followed by reconstruction, (i) regional maxima operation applied on opening–closing by reconstruction, (j) regional maxima superimposed on the original image, (k) modified regional maxima superimposed on the original image, (l) thresholded operation applied on image opening–closing by reconstructed image, (m) watershed ridgelines, (n) markers and object boundaries superimposed on the original image, (o) colored watershed image, and (p) color watershed image superimposed transparently on the original image.

reconstruction in which the maximum pixel value of 255 is applied and in other regions, the original image pixel value is retained. The gray level thresholding is applied on the opening–closing by reconstruction to obtain the binary image. Then, watershed transform is applied to the binary image to yield the watershed rigid lines. But here, the watershed rigid lines are not clear and appear as a black image. Then, the markers and object boundaries are superimposed on the original thermal image. Finally, the colored watershed segmented output is obtained in which each color represents the specific region. The green color region indicates the segmented ROI;

blue color represents the other regions; and the regions with yellow, orange, and mustard colors depict the background region with variational intensities.

Figure 5.8 represents the morphology-based watershed segmentation algorithm implemented on the hand thermal image. Figure 5.8a shows the original input thermal image of RA patients, Figure 5.8b indicates the thermal image converted into grayscale image, Figure 5.8c depicts the gradient magnitude image, Figure 5.8d displays the watershed transform of gradient magnitude image, Figure 5.8e portrays the image opening operation applied on the transformed image, Figure 5.8f shows the image opening operator followed by the application of image reconstruction operation. Figure 5.8g represents the combination of image opening and closing operation applied on the image, Figure 5.8h depicts the image opening–closing operation followed by reconstruction, Figure 5.8i indicates the regional maxima operation applied on opening–closing by reconstruction, Figure 5.8j displays the regional maxima superimposed on the original image, Figure 5.8k represents the modified regional maxima superimposed on the original image, Figure 5.8l indicates the thresholded operation applied on image opening–closing by reconstructed image, Figure 5.8m depicts the watershed ridgelines, Figure 5.8n shows the markers and object boundaries superimposed on the original image, Figure 5.8o represents the colored watershed image, and Figure 5.8p indicates the color watershed image superimposed transparently on the original image.

5.6 STATISTICAL FEATURE EXTRACTION AND CLASSIFICATION

5.6.1 ORIENTED FAST AND ROTATED BRIEF

ORB is a fast and accurate orientation keypoint detector and descriptor. The ORB feature is extracted based on the combination of the detection method FAST and BRIEF descriptors. ORB is considered as the alternative method and performs faster than SIFT and SURF. It has a fast computation speed mechanism and high matching accuracy.

The ORB algorithm used FAST algorithm to detect feature points and consists of the following processes: (i) feature extraction; (ii) feature descriptors, and (iii) feature matching process.

The FAST corner point obtained has a large number of feature points but information about the direction was not found. Hence, ORB algorithms were improved further by computing of Harris response value for the FAST corner point based on the gray value, and the first points were taken into consideration as feature points.

The feature point direction is estimated by using the intensity centroid method to extract the feature points that have the property of rotational invariance. The ORB used modified BRIEF algorithm for obtaining the feature descriptors for each point. The similarity between the feature point descriptor to find the matching in two different images is important to be considered in feature point matching (Luo et al., 2019). Brute force matcher is the method used for calculating the similarity between feature points in two different images based on distance. Hamming distance method is preferred in the case of binary BRIEF descriptors.

5.6.2 SUPPORT VECTOR MACHINE

Support vector machine is a supervised classification method used for binary classification problems and regression.

Create the hyperplane and create another two margin lines having some distance which is linearly separable. After creating the hyperplane, the dotted hyperplane parallel to the main hyperplane passes through the nearest positive points D+, and another hyperplane parallel to the main hyperplane passes through the nearest negative points D−. The distance between the two parallel lines is called the margin.

Apart from hyperplane, we can create multiple hyperplanes, but we need to focus on the marginal plane. Maximize the particular marginal plane or marginal distance particularly applicable for linearly separable points.

Support vectors are the nearest positive points or negative points near the margin of the hyperplanes. The equation of hyperplane and optimization function of SVM classifier is given as below from equation 5.5 to 5.10:

Equation of hyperplane:

$$\omega^T x + b = 0 \tag{5.5}$$

Negative plane:

$$\omega^T x + b = -1 \tag{5.6}$$

Positive plane:

$$\omega^T x + b = 1 \tag{5.7}$$

Solving Equations (5.6) and (5.7), we get

$$\omega^T \left(x_2 - x_1 \right) = 2 \tag{5.8}$$

Remove ωT by taking $\| \omega \|$

$$\frac{\omega T \left(x2 - x1 \right)}{\|\omega\|} = \frac{2}{\|\omega\|} \tag{5.9}$$

$$\frac{\left(x2 - x1 \right)}{\|\omega\|} = \frac{2}{\|\omega\|}$$

The optimization function is the maximum of $\dfrac{2}{\|\omega\|}$ such that

$$y_i = \omega^T x_i + b_i \geq 1$$

To optimize my model

$$\left(\omega * b *\right) = \min \frac{\|\omega\|}{2} + c_i \sum_{i=1}^{n} \varepsilon_i \qquad (5.10)$$

where c_i represents how many errors the model can consider and ε_i indicates the value of the error.

5.6.3 k-NEAREST NEIGHBOR

Suppose we have 2D data set with two classification problems, the k-nearest neighbor algorithm function is as follows:

- Select k-value and choose the number of k-nearest neighbors.
- For example, if $k = 5$, the nearest 5 neighbors in the distributed data is considered.
- Calculate the distance of the nearest neighbors for $k = 5$.
- Find out the number of nearest neighbors belonging to Category 1 and Category 2.
- If the maximum nearest neighbor is from Category 1, the query point belongs to Category 1.
- Then, the distance between the query point and nearest neighbors is calculated using Euclidean distance or manhattan distances.

If the dataset is completely imbalanced, k-NN is biased with respect to output. As the k-value increases, the error rate comes down and then increases. After standard scale to the data and finding how the data is distributed, assume k-values (e.g., say in the range from 1 to 40), plot the accuracy rate, and check the accuracy rate for stability. Then, 10-fold cross-validation is used to calculate the error rate and plot the error rate. Find the k-value. The error rate is gradually reduced; the k-value goes down. The accuracy rate is continuously stable and good at a particular k-value.

5.6.4 LINEAR DISCRIMINANT ANALYSIS (LDA)

Linear discriminant analysis is a supervised classification technique mainly used for dimensionality reduction of data sets. It is used for the classification of features between two classes and is applicable for multi-class problems. LDA hypothesizes based on the assumption that all classes as linearly separable with multiple discrimination functions indicating several hyperplanes in feature space to create distinct classes. If the classification problem is based on two classes, then LDA uses one hyperplane and projects the features into this hyperplane based on the following criteria: (i) computation of mean of two classes and maximizing the distance between them and (ii) minimizing the variation between each class.

To illustrate the concepts of LDA in a detailed manner, let us consider the two-class dataset: class A as Normal and class B as RA.

The prior probabilities of class A and class B are P1 and P2, respectively. Means of class A and class B are μ_A and μ_B, respectively.

$$\text{Total mean } \mu = P1 \times \mu_A + P2 \times \mu_B \qquad (5.11)$$

The within-class scatter matrix is computed as

$$Sw = \sum_{j=1}^{c} Pj \times \text{cov } j$$

For the two-class problem,

$$Sw = \sum_{j=1}^{2} Pj \times \text{cov } j \qquad (5.12)$$

$$\text{cov}_j = \left(xj - \mu j \right)\left(xj - \mu j \right)^{\mathrm{T}} \qquad (5.13)$$

The between-class scatter matrix is

$$S_b = 1/c \sum_{j=1}^{c} \left(\mu j - \mu \right)\left(\mu j - \mu \right) T$$

$$= 1/c \sum_{j=1}^{2} \left(\mu j - \mu \right)\left(\mu j - \mu \right) T \qquad (5.14)$$

The discriminant plane is obtained to find the maximum ratio between the between-class scatter variances and within-class scatters provided in Equation (5.15).

$$\max \left\{ Jlda \right\} = \frac{w S_b w^{T}}{w S_w w^{T}} \qquad (5.15)$$

5.6.5 Random Forest Classifiers

Random forest is a supervised machine learning algorithm that performs both classification and regression. It constructs decision trees on randomly chosen data sets and functions by the divide-and-conquer approach. The prediction from the individual decision tree is based on attribute selectors such as information gain, Gini index, and gain ratio. The best solution from each decision tree was obtained by means of a voting process. The decision tree which produces majority votes is considered as the final prediction. The advantages of RF classifiers are that they are a highly accurate and robust method due to the maximum number of decision

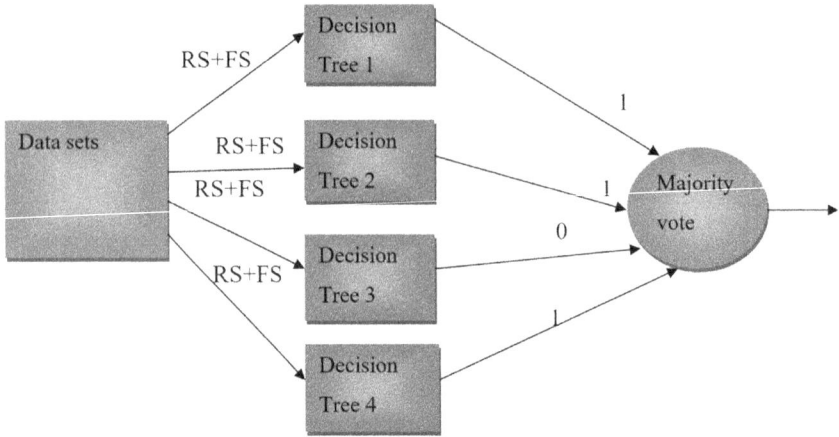

FIGURE 5.9 Block diagram of random forest classifier.

trees involved in the process, avoid overfitting problems, and can handle missing values, e.g., replacing median values in the continuous variables.

The block diagram of the random forest classifier is illustrated in Figure 5.9. Consider the dataset containing N number of samples; row sampling and feature sampling were performed and given as an input to each decision tree. While doing this process, some of the records and data get repeated. The decision tree has two main properties such as low bias and high variance. Low bias means training error is less and high variances indicate that the data is prone to large variance. In RF classifier, multiple decision trees are created, based on the majority vote provided by all the decision trees, and high variance of data is converted into low variance. Tuning the hyperparameter is one of the important criteria that needs to be considered in the random forest classifier. The hyperparameters are chosen for the RF classifier based on the number of decision trees in the forest and the number of features selected by the decision tree when splitting a node.

Optimization of hyperparameter is performed by using cross-validation method to avoid overfitting process. In machine learning problems, we split data into training and testing sets. We split our training data into k number of subsets called folds.

Consider $k = 10$. In the first iteration, we train on the first 9 folds and evaluate on the 10th. In the second iteration, we train on the first, second, third, etc., and evaluate on the 9th fold and repeat the procedure eight more times, each time evaluating on a different fold. At the end of the training process, the performance of each fold is averaged to yield final validation metrics.

5.6.6 Naïve Bayes Classifier

The Naïve Bayes algorithm is basically a supervised classification technique based on conditional probabilities on the given data.

$$P\left(\frac{A}{B}\right) = \frac{P\left(\frac{B}{A}\right)P(A)}{P(B)} \qquad (5.16)$$

Consider a dependent data set $X = \{x_1, x_2, x_3, x_4 \ldots x_n\}$ and output feature $\{y_1, y_2, y_3, y_4, \ldots y_n\}$.

The probability of Y given X is

$$P\left\{\frac{Y}{x1}, x2, x3, x4 \ldots xn\right\}$$

$$= \frac{P\left(\frac{x1}{y}\right)P\left(\frac{x2}{y}\right)P\left(\frac{x3}{y}\right)\ldots.P\left(\frac{xn}{y}\right)P(y)}{p(x1)\,p(x2)\,p(x3)\ldots.p(xn)} \qquad (5.17)$$

$$= \frac{p(y)\prod_{i=1}^{n}p\left(\frac{xi}{y}\right)}{p(x1)\,p(x2)\,p(x3)\ldots.p(xn)}$$

$$P\left\{\frac{Y}{x1}, x2, x3, x4 \ldots xn\right\}\alpha\; p(y)\prod_{i=1}^{n}p\left(\frac{xi}{y}\right) \qquad (5.18)$$

$$y = \text{argmax } y\left\{p(y)\prod_{i=1}^{n}p\left(\frac{xi}{y}\right)\right\} \qquad (5.19)$$

To estimate $p(y)$ and $p(xi/y)$, maximum A posterior (MAP) is used, where $p(y)$ is the relative frequency of occurrence of class y in the training set. The advantages of Naïve Bayes classifier were easy to implement, fast, and real-time predictions of class of test data set.

5.7 FINDINGS AND CONCLUSION

The proposed study demonstrated the efficacy of thermal imaging to be used as a pre-screening tool for the detection of RA. Quantitative parameters such as HDI and thermographic index and its grading system to assess the disease severity of RA were discussed. The various image segmentation algorithms were used for analyzing the thermal images of RA and normal subjects. Compared to the manual thresholding method, the performance of the automated thresholding method was better in segmenting the thermal images of RA patients. But since the thresholding is based on intensity, apart from segmenting the abnormal ROI, it segments the other high-intensity regions as shown in Figure 5.4d. Also, another drawback of the

TABLE 5.1
Performance Metrics for RA Detection Based on Machine Learning Classifiers

Classifier	Precision	F1-Score	Recall	Accuracy	AUC
SVM-RBF kernel	1.0000	0.9862	0.9733	0.9867	1.0000
SVM-Linear Kernel	0.9765	0.9737	0.9733	0.9771	0.9993
k-NN	1.00	0.99	0.97	0.99	1.00
LDA	1.00	0.9931	0.9867	0.9933	1.00
Random forest	1.00	0.99	0.99	0.99	1.00
Naïve Bayes	0.93	0.96	1.00	0.97	1.00

automated thresholding method was it holds good and produces better segmentation results only when the background is free from noise. The conventional edge detection operators applied on thermal image produces only the vertical and horizontal edges present in the image. But sometimes, detection of false edges is possible and it is sensitive to noise. When compared to other intensity- and discontinuities-based segmentation methods, clustering-based approach performed well in segmenting the thermal images of RA patients. Hence, k-means segmentation method applied in thermal images could segment the abnormal hotspot ROI in the thermal images of RA patients. The watershed segmentation method performs the operation by a combination of discontinuities- and similarity-based approaches. This method differentiates the region based on color and intensity-based method. The disadvantage of watershed transformation involved excessive over-segmentation of images. It fails to provide proper results in case of improper background and the images are prone to noise. Hence, k-means segmentation method is adopted in this proposed study for the segmentation of abnormal ROIs, and ORB feature extraction method is used for the extraction of features followed by various machine learning classifiers.

Among the implemented machine learning classifiers, k-NN, random forest, and LDA produced good accuracy of 99% compared to others in the classification of normal and RA patients. The precision score was highest for SVM-RBF kernel, k-NN, random forest, and LDA classifiers compared to other classifiers. A 5-fold cross-validation method is employed in the training and validation of classifiers.

In conclusion, among the various segmentation methods applied in thermal images of RA patients, the k-means clustering technique provided better segmentation results compared to other methods. ORB feature extraction method in combination with various machine learning classifiers produced very good accuracy in the classification of normal and RA patients. Table 5.1 shows results based on ORB feature extractor and machine learning classifiers.

5.8   CASE STUDY

RA is an inflammatory multi-factorial systemic disorder that affects around 1–2% of Indians, especially women in the age group of 50-year-olds (Mittal and Dubey,

2013). It causes pain, disability, and loss of function. Earlier screening or diagnosis of RA allows better treatment and prevention. Unfortunately, there is no specific test to detect the presence of RA. RA is traditionally diagnosed by combining symptoms such as morning stiffness with clinical signs such as a number of swollen and tender joints and laboratory tests such as ESR, CRP, rheumatoid factor, and anti-cyclic citrullinated peptide (anti-CCP) antibodies. For the diagnosis of RA, clinical signs and symptoms must have been present for at least 6 weeks (Arnett et al., 1988). The severity of disease can be further assessed by imaging modalities such as X-ray, ultrasound, and MRI. However, the sensitivity of these tests is limited in the diagnosis of RA at the earlier stage (Heidari, 2011). In 2010, the American College of Rheumatology/European League against Rheumatism Collaborative Initiative published revised rheumatoid arthritis classification criteria aimed at earliest identification of patients who might benefit from treatment with disease-modifying anti-rheumatic drugs (DMARD). The presence of obvious "clinical" synovitis in at least one joint is central to this classification (Aletaha et al., 2010). Imaging techniques such as MRI or sonography can confirm clinically suspected synovitis.

According to evidence-based recommendations of the European League against Rheumatism (EULAR) task force (Colebatch et al., 2012), there is a need for the development of non-invasive and cost-effective diagnostic techniques for RA, particularly in the early stage. Moreover, the earlier the diagnosis of RA, the earlier the patient receives treatment. Thus, earlier diagnosis is crucial in the treatment of RA. MRI is commonly used in the identification of RA in the early stage. However, its utility is limited by its inherent cost factor. On the other hand, thermal imaging modality and ultrasound play a vital role in the early assessment of RA due to it being cost-effective and easily accessible. The metabolic activity in rheumatoid inflammation is very complex and is not correctly described as metabolism on a high level. The amount of heat released by inflammation remains undefined. Also, the transfer of inflammatory heat from the joint to the skin requires further research.

The major signs of RA are inflammation and synovial hypertrophy and are associated with pain and structural damage caused by the disease (Roemer et al., 2009). Hence, color Doppler ultrasound (CDUS) can be used to quantify intrasynovial blood flow to assess synovial membrane inflammation. In addition, the EULAR task force recommended the importance of ultrasound in the evaluation of RA.

In this context, this case study demonstrates the role of non-invasive and cost-effective techniques, namely IR thermography and CDUS in the diagnosis of earlier stage RA. The aim of this case study was to correlate thermographic and color Doppler ultrasonographic knee features in RA patients using clinical and functional assessment as the gold standard. This study provides an in-depth insight into the various steps involved in the thermographic and CDUS evaluation of RA. The following three case studies demonstrate the potential diagnostic value of thermography and CDUS in the evaluation of rheumatoid arthritis, and how they can change the present clinical management.

5.8.1 COLOR DOPPLER ULTRASONOGRAPHIC EVALUATION

Color Doppler ultrasonographic examination is performed after clinical and thermographic investigation. Conventional grayscale US is used to investigate the knee joint starting at lateral joint space, then to lateral, medial, central, and suprapatellar recess, and medial joint space. Both longitudinal and transverse US images were taken.

The CDUS parameters were elaborated as follows:

1. **Knee Joint Effusion**: It is the maximum AP diameter measured in the supine position of suprapatellar recess with extended knee joint and contracted biceps femoris muscle.
   Semi-quantitative grading was used to measure the effect of effusion based on Fiocco et al. (1996) as follows: (i) Grade 0 is assigned for effusion < 2 mm; (ii) Grade 1 is considered for effusion varies from 2 to 5 mm; (iii) Grade 2 indicates the effusion level is in the range of 6–8 mm; (iv) Grade 3 defines the effusion level greater than 8 mm.
2. **Echogenity of Effusion**: The echos of effusion is subjectively graded using a 0–3 scale indicated in previous studies (Schmidt et al., 2000).
   Grade 0: no echos in the effusion
   Grade 1: mild echos
   Grade 2: moderate echos
3. Perfusion Intensity on CDUS Images

The perfusion intensity is determined by maximum color Doppler activity observed in ultrasound images for each patient. The semi-quantitative grading scale in the range of 0–3 was used to study the perfusion level based on the study conducted by Schmidt et al. (2000).

Grade 0: absence of color Doppler signal (no perfusion)
Grade 1: 1–3 color pixels (mild perfusion)
Grade 2: 3–10 color pixels (moderate perfusion)
Grade 3: >10 color pixels (high perfusion)

The perfusion intensity is analyzed using ImageJ software (Wayne Rasband, 1997). ImageJ is used to analyze the CDUS images to determine the intensity of perfusion (Denoble et al., 2010).

The steps involved in segmenting the color pixels from CDUS images are as follows:

1. **Selection of ROI**: After the US image is loaded into the ImageJ platform, the ROI is selected using the in-built function available in ImageJ. The ROI is traced in the synovium tissue which is characterized by the hypoechoic mass near the joint covering the bony surfaces. The image is then cropped to remove the unwanted regions.

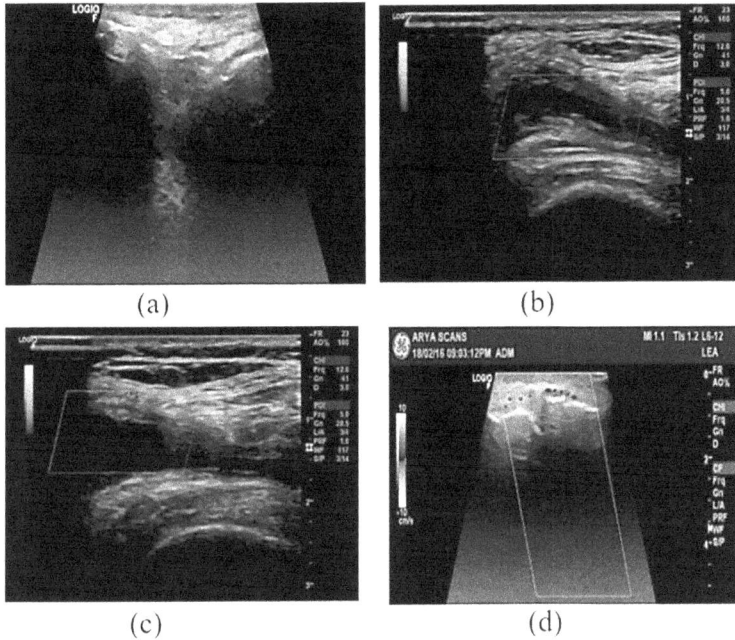

FIGURE 5.10 CDUS image of the control subject with no knee joint effusion. (b) CDUS image showing Grade 1 knee joint perfusion (RA subjects). (c) CDUS image showing Grade 2 knee joint perfusion (RA subjects). (d) CDUS image of severe perfusion (Grade 3).

2. **Smoothing:** Smoothing is done to eliminate the problems caused by the cell boundaries.
3. **RGB Color Transformation:** The image is split into R, G, B color channels, and the image with the best contrast is selected.
4. **Thresholding:** The color pixels are segmented using autothreshold method given in the ImageJ toolkit.

Figure 5.10 shows the CDUS image of the control subject, with no signs of knee joint effusion and perfusion. In the control subjects, there are no knee joint effusions and perfusions: (a) no perfusion (Grade 0), (b) mild perfusion (Grade 1), (c) moderate perfusion (Grade 2), and (d) severe perfusion (Grade 3).

5.8.2 CASE REPORTS

5.8.2.1 Case Study 1
This case study deals with a 25-year-old male who presented with RA for more than 1 year. The subject complained of persistent swelling, numbness, pain, and stiffness in the knee, feet, and wrist region. During the visit, the subject marked

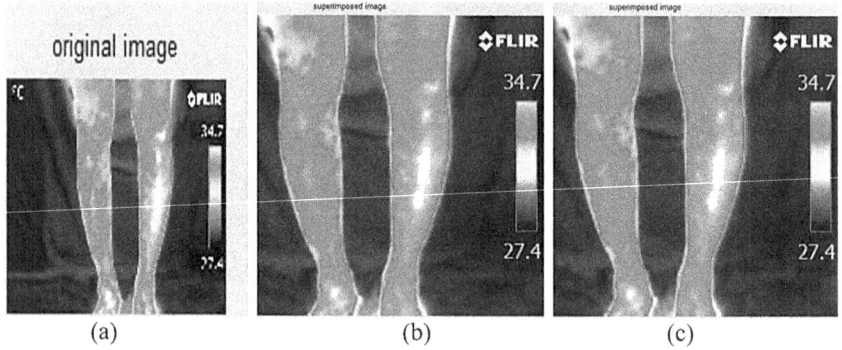

FIGURE 5.11   Thermal image of the knee region of RA patients for Case Study 1: (a) input thermal image, (b) segmented output image, and (c) segmented image superimposed with the input image.

the pain on VAS as 8 out of 10 (0 indicates no pain and 10 indicates severe pain). The Westergren results showed positive signs of RA (22 mm/h). ESR of the patient was slightly elevated (normal value for men under 30 years <15 mm/h). However, ESR is only diagnostic for RA in combination with proven clinical synovitis. The subject reported a family history of RA. His father and grandfather reported RA during their 50 s. The subject has been under medications for more than 8 months, with moderate relief. He was examined using both IR thermography and CDUS. The thermographic results showed minor hot spots near the knee region possibly indicating synovitis, and a large hot area in the middle of the lower leg not associated with any articular structure (Figure 5.11). The mean temperature obtained from the knee region was found to be 34.8°C. The results of CDUS showed the presence of knee joint effusion with no signs of perfusion. The maximum AP diameter of the suprapatellar recess is 1.5 mm which shows Grade 1 knee joint effusion as indicated in Figure 5.8b.

5.8.2.2   Case Study 2

The second case study deals with a 48-year-old woman who presented with RA. The subject reported that the disease developed about 6 months earlier, with pain in the knee region. The pain then extended to all joints of the body, namely fingers, wrists, shoulder, and ankle joints. The subject was a homemaker and was not able to do daily activities. No systemic condition like diabetes mellitus or cardiac problem has been reported. Westergren obtained a slightly elevated ESR (30 mm/h). During the visit, the subject marked the pain on VAS as 8 out of 10 (0 indicates no pain and 10 indicates severe pain). No family history of RA has been reported. The subject was under treatment and her symptoms subsided gradually. The patient was evaluated with both thermography and CDUS. Figure 5.12 represents the thermography results of the patient showing a hot spot in the right leg, which is not related to an articular structure. The mean temperature obtained from the knee region was found to be 33.9°C. Similarly, the CDUS image of the patient

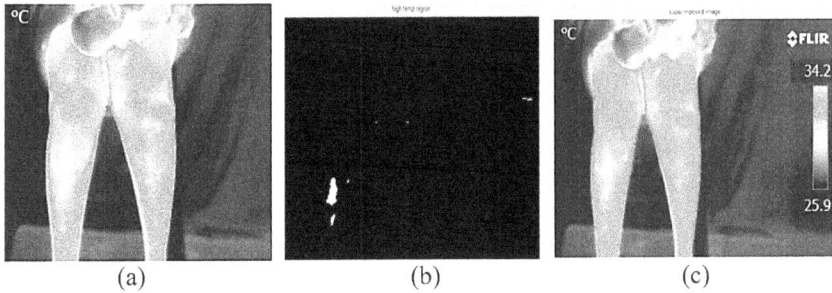

(a)   (b)   (c)

FIGURE 5.12 Thermal image of the knee region of RA patients for case Study 2: (a) input thermal image, (b) segmented output image, and (c) segmented image superimposed with the input image.

is given in Figure 5.12c as Grade 2 perfusion. CDUS findings revealed knee joint effusion with AP diameter ranging from 3 mm. The echogenity is graded by the sonographer.

5.8.2.3 Case Study 3

The third case study deals with a 45-year-old man who reported RA which had developed 6 months earlier.

IR Thermal Analysis: The patient was asked to sit in a temperature-controlled room and a thermal image was obtained from the subjects as per the above protocol. The mean temperature obtained from the knee region was found to be 34.3°C. During the visit, the subject reported pain using VAS as 8/10, where 0 = no pain and 10 = severe pain, and reported health status as 6/10, where 0 = no pain and 10 = severe pain. Laboratory evaluation indicated elevated levels of ESR in the range of 30 mm/h. The subject did not respond to medication and reported worsening of symptoms with the onset of winter. The patient was diagnosed with thermography as well as CDUS. Figure 5.13 represents the thermography results of the patient showing a hot spot in the leg. Similarly, the CDUS image of the patient is given in Figure 5.13d as Grade 3 perfusion. CDUS findings revealed knee joint effusion with AP diameter ranging from 6.3 mm. The echogenity is graded by the sonographer. Figure 5.13a shows the thermal image of the knee region of the RA patient and Figure 5.13b represents the segmented hot spot region. Figure 5.13c indicates the segmented image superimposed on the original image.

5.8.3 Results of CDUS Examination

Table 5.2 reveals the color Doppler findings of healthy control subjects and five patients with rheumatoid arthritis at the knees. Out of the five RA patients, knee joint effusion and perfusion are found only in four patients. Likewise, two control subjects do not reveal any signs of knee joint effusion and perfusion. A mild degree of knee joint effusion and perfusion was found in three images out of four

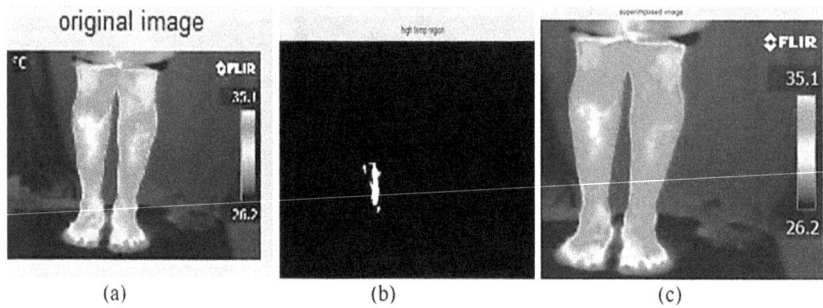

original image

(a)   (b)   (c)

FIGURE 5.13 Thermal image of knee region of RA patients: (a) input thermal image, (b) segmented output image, and (c) segmented image superimposed with the input image.

TABLE 5.2
Color Doppler Findings of RA Patient

Patient No.	Extent of Effusion	Intensity of Perfusion	Anterior–Posterior Diameter (mm)
1	2	2	6.3
2	1	1	2.05
3	0	0	1.5
4	1	1	3.0
5	0	0	0

RA positive CDUS images. Only one RA subject revealed a moderate degree of perfusion and knee joint effusion in the CDUS findings.

5.8.4 Comparison of Clinical, Thermographic, and CDUS Findings

Table 5.3 compares the results of clinical, thermographic, and CDUS findings of five RA patients. On the other hand, out of the five CDUS images, the results of four subjects correlated with clinical findings. The four CDUS images provided positive findings for RA. However, only one CDUS image does not correlate with the clinical findings.

5.8.5 Discussion

The proposed case studies focused on the comparison of clinical, thermographic, and CDUS parameters of RA on case to case basis. The severity of RA has been identified in terms of clinical parameters like ESR, skin surface temperature measurement, and CDUS parameters such as effusion and perfusion. We have compared the thermography and CDUS findings by correlating the clinical findings.

The clinical and functional assessment score correlates with IR thermal image analysis more effectively than CDUS. Out of the five CDUS images, the results of

TABLE 5.3
Comparison of Clinical, Thermal, and Ultrasound Parameters for Different Subjects

Patient No.	ESR	Skin Surface Temperature (°C)	CDUS Perfusion
1	22	34.8	Grade 1
2	23	34.3	Grade 1
3	28	33.9	Grade 1
4	30	34.4	Grade 2
5	25	33.2	Grade 0

four images are comparable to the clinical findings. However, the result of one CDUS image does not correlate with clinical findings. However, the proposed study results contradict the earlier studies (Beitinger et al., 2013, Lasanen et al., 2015) in which Doppler ultrasound was efficient in determining synovitis and effusions. Similarly, in the study by Qvistgarrd et al. (2001) using 18 RA joints, the vascularisation parameter obtained by CDUS correlated moderately, but correlated significantly with ESR scores. Hence, no definite conclusion can be drawn from this study, as the results of the ultrasound image analysis are operator- as well as instrument-dependent. Hence, the present study warrants further investigation with a large sample size of data.

The results obtained from the case study indicated that both IR thermal imaging and CDUS offer good diagnostic potential in detecting RA pathology. Future studies should compare these two diagnostic techniques using a large sample size and appropriate study design.

5.8.6 LIMITATIONS AND RECOMMENDATIONS

1. The study design limits the results of the present study because neither criteria for recruiting patients and controls nor clear outcomes have been defined. Since the sonographers were not blinded to the clinical information; there would have been a certain observer bias. Hence, randomized blinded control study should be adopted in future to eliminate observer bias.
2. CDUS examination is subjected to artifacts and false positives, hence standardization of CDUS is necessary. Hence, future studies should focus on the standardization of CDUS.
3. The performance of thermography is affected by body and room temperature, physical activity, food habits prior to the study, and also by the lack of standards for camera views, position of the patient, and analyzing thermal images. Since all these factors can influence the results of the thermal findings, in future, measures should be taken to consider these factors.

REFERENCES

Aletaha D, Neogi T, Silman AJ, Funovits J, Felson DT, Bingham III CO et al. 2010 rheumatoid arthritis classification criteria: An American College of Rheumatology/European League Against Rheumatism collaborative initiative. *Arthritis & Rheumatology* 2010, 62(9): 2569–2581.

Ammer K and Ring F. *The Thermal Human Body*. A Practical Guide to Thermal Imaging. Jenny Standord Publishing, Singapore, 2019.

Anderson JK, Zimmerman L, Caplan L, Michaud K Measures of rheumatoid arthritis disease activity. *Arthritis Care and Research* 2011, 63(S119): S14–S36.

Arnett FC, Edworthy SM, Bloch DA, McShane DJ, Fries JF, Cooper NS, et al. The American Rheumatism Association 1987 revised criteria for the classification of rheumatoid arthritis. *Arthritis & Rheumatology* 1988, 31: 315–324.

Beitinger N, Ehrenstein B, Schreiner B, Fleck M, Grifka J, Lüring C, Hartung W et al. The value of colour doppler sonography of the knee joint: A useful tool to discriminate inflammatory from non-inflammatory disease? *Rheumatology* 2013, 52(8): 1425–1428.

Chaudhary AS, Chaturvedi DK Efficient thermal image segmentation for heat visualization in solar panels and batteries using watershed transform I.J. *International Journal of Image, Graphics and Signal Processing* 2017, 11: 10–17.

Chen W, Cheng H, Shen H An effective methodology for thermal characterization of electronic packaging. *IEEE Transactions on Components, Packaging and Manufacturing Technology* 2003, 26: 222–232.

Colebatch AN, et al. EULAR recommendations for the use of imaging of the joints in the clinical management of rheumatoid arthritis. *Annals of the Rheumatic Diseases* 2012, 72(6): 804–814. doi:10.1136/annrheumdis-2012-203158

Collins AJ, Ring EFJ, Cosh JA, Bacon PA Quantitation of thermography in arthritis using multi-isothermal analysis. *Annals of the Rheumatic Diseases Disorders* 1974, 33: 113–115.

Denoble AE, Hall N, Pieper CF, Kraus VB Patellar skin surface temperature by thermography reflects knee osteoarthritis severity. *Clinical Medicine Insights: Arthritis and Musculoskeletal Disorders* 2010, 3: 69–75.

Devereaux MD, Parr GR, Lachmann SM, Page Thomas DP, Hazleman BL Thermographic diagnosis in athletes with patellofemoral Arthralgia. *Journal of Bone and Joint Surgery* 1986, 68(1): 42–44.

Devereaux MD, Parr GR, Page Thomas DP, Hazleman BL Disease activity indexes in rheumatoid arthritis; a prospective comparative study with thermography. *Annals of the Rheumatic Diseases Disorders* 1985, 44: 434–437.

Duarte A, Carrão L, Espanha M, Viana T, Freitas D, Bártolo P, Faria P, Almeida HA Segmentation algorithms for thermal images. *Procedia Technology* 2014, 16: 1560–1569.

Fiocco U, Cozzi L, Rubaltelli L, et al. Long-term sonographic follow-up of rheumatoid and psoriatic proliferative knee joint synovitis. *British Society for Rheumatology* 1996, 35: 155–163.

Fries J, Spitz PW, Young DY The dimensions of health outcomes: The health assessment questionnaire, disability and pain scales. *The Journal of Rheumatology* 1982, 9: 789–793.

Frize M, Adea C, Payeur P, Di Primio G, Karsh J, Ogungbemile A Detection of rheumatoid arthritis using infrared imaging. *Proceedings of SPIE* 2011, 7962.

Ghosh A, Ghosh B, Pain S, Pande A, Saha S, Banarjee A, Biswas A Comparison between DAS28, CDAI and HAQ-DI as tools to monitor early rheumatoid arthritis patients in eastern India. *International Journal of Rheumatology* 2011, 6(3): 116–122.

Gonzalez RC, Woods RE *Digital Image Processing* 3'd Edition, Prentice Hall, 2007, 598–603.

Heidari B Rheumatoid arthritis: Early diagnosis and treatment outcomes. *Caspian Journal of Internal Medicine* winter; 2011, 2(1): 161–170.

Wayne Rasband, Image Processing and Analysis in Java (ImageJ) Software Documentation, 1997, http://rsbweb.nih.gov/ij/

Jadin M Recent progress in diagnosing the reliability of electrical equipment. *Infrared Physics and Technology* 2012, 55(4): 236–245.

Kacmaz S, Ercelebi E, Zengin S, Cindoruk S The use of infrared thermal imaging in the diagnosis of deep vein thrombosis. *Infrared Physics and Technology* 2017, 86: 120–129.

Kadir T, Brady M Saliency, scale and image description. *International Journal of Computer Vision* 2001, 45: 83–105.

Krishnan E, Sokka T, Häkkinen A, Hubert H, Hannonen P Normative values for the health assessment questionnaire disability index. *Arthritis & Rheumatology* 2004, 50: 953–960.

Lasanen R, et al., Thermal imaging in screening of joint inflammation and rheumatoid arthritis in children. *Physiological Measurement* 2015, 36: 273.

Likas A, Vlassis N, Verbeek JJ The global k-means clustering algorithm. *Pattern Recognition* 2003, 36(2): 451–461. doi:10.1016/s0031-3203(02)00060-2

Luo C, Yang W, Huang P, Zhou J Overview of image matching based on ORB algorithm. *IOP Conference Series* 2019, 1237: 032020.

Maeda M, Kachi H, Ichihashi N, Oyama Z, Kitajima Y The effect of electrical acupuncture stimulation therapy using thermography and plasma endothelin levels in patients with progressive systemic sclerosis. *Journal of Dermatology Science* 1998, 17(2): 151–155.

Mishra R, Sharma BL, Gupta R, Pandya S, Agarwal S, Agarwal P, Grover S, Sarma P, Wanjam K Indian Rheumatology Association consensus statement on the management of adults with rheumatoid arthritis. *Indian Journal of Rheumatology* 2008, 3(3): S1–S16.

Mittal A, Dubey SK Analysis of MRI images of Rheumatoid Arthritis through morphological image processing techniques. *International Journal of Computing Science* 2013, 10(3): 118–122.

Nameirakpam D, Khumanthem M, Yambem JC Image segmentation using K-means clustering algorithm and subtractive clustering algorithm. *11th International Conference on Signal, Image Processing and Pattern Recognition*, 2015, 54: 764–771, August.

Nandagopan GL, Haripriya AB Implementation and comparison of two image segmentation techniques on thermal foot images and detection of ulceration using asymmetry. *2016 International Conference on Communication and Signal Processing (ICCSP)*, Melmaruvathur, India, 2016: 0356–0360.

Nandhakumar N, Aggarwal JK Integrated analysis of thermal and visual images for scene interpretation. *IEEE Transaction on Pattern Analysis and Machine Intelligence* 1988, 10(4): 469–481.

Online disease activity score calculator for rheumatoid arthritis, available from http://www.das-score.nl. Accessed on 2021.

Preetha MMSJ, Suresh LP, Bosco MJ Image segmentation using seeded region growing. *IEEE International Conference on Computing, Electronics and Electrical Technologies (ICCEET)* 2012, 576–583.

Qvistgarrd E, Rogind H, Torp-Pedersen S, et al. Quantitative ultrasonography in rheumatoid arthritis: Evaluation of inflammation by doppler technique. *Annals of the Rheumatic Diseases* 2001, 60: 690–693.

Ramírez-Rozo TJ, García-Álvarez JC, Castellanos-Domínguez CG Infrared thermal image segmentation using expectation-maximization-based clustering. *2012 XVII Symposium of Image Signal Processing, and Artificial Vision (STSIVA)*, 2012, 223–226. doi: 10.1109/STSIVA.2012.6340586

Ring EFJ Quantitative thermography and thermographic index. *Verh Dtsch Ges Rheumatology* 1980, 6: 287–288.

Roemer FW, Zhang Y, Niu J et al. Tibiofemoral joint osteoarthritis: Risk factors for MR-depicted fast cartilage loss over a 30-month period in the multicenter osteoarthritis study. *Radiology* 2009, 252: 772–780.

Salisbury R, Parr G, De Silva M, Hazleman BL, Page-Thomas DP Heat distribution over normal and abnormal joints: thermal pattern and quantification. *Annals of the Rheumatic Diseases* 1983, 42(5): 494–499.

Schmidt WA, Völker L, Zacher J, Schläfke M, Ruhnke M, Gromnica-Ihle E Colour doppler ultrasonography to detect pannus in knee joint synovitis. *Clinical and Experimental Rheumatology* 2000, 2000(18): 439–444.

Sergyan S *Edge detection techniques of thermal images IEEE 10th Jubilee International Symposium on Intelligent Systems and Informatics Subotica*, Serbia 2012.

Shaikh S, Gite H, Manza RR, Kale KV, Akhter N Segmentation of thermal images using thresholding-based methods for detection of malignant tumours. *Intelligent Systems: Technology and Applications* 2016, 2016: 131–146.

Snekhalatha U, Anburajan M, Venkatraman B, Menaka M, Raj B Evaluation of rheumatoid arthritis in small animal model using thermal imaging. *Proceedings of the International Conference Signal Processing Communication Computing and Networking Technologies* 2011, 785–791.

Snekhalatha U, Anburajan M, Sowmiya V, Venkatraman B, Menaka M Automated hand thermal image segmentation and feature extraction in the evaluation of rheumatoid arthritis. *Proceedings of the Institution of Mechanical Engineers, Part H: Journal of Engineering in Medicine* 2015, 229(4): 319–331.

Snekhalatha U, Sowmiya V, Nilkantha G A computer aided diagnostic based Thermal image Analysis: A potential tool for the Evaluation of Rheumatoid Arthritis in Hand. *Journal of Medical and Biological Engineering* 2017, 38(4): 666–677.

Spalding SJ, Kwoh CK, Boudreau R, Enama J, Lunich J, Huber D, Denes L, Hirsch R Three dimensional and thermal surface imaging produces reliable measures of joint shape and temperature: A potential tool for quantifying arthritis. *Arthritis Research & Therapy* 2008, 10(1): R10. doi:10.1186/ar2360

Suma AB, Snekhalatha U, Rajalakshmi T Automated thermal image segmentation of knee rheumatoid arthritis. *2016 International Conference on Communication and Signal Processing (ICCSP)* (2016), pp. 0535–0539.

Wang W, Zhang J, Shen C Improved human detection and classification in thermal images. *IEEE International Conference on Image Processing* 2010, 2313–2316.

Warashina H, Hasegawa Y, Tsuchiya H, Kitamura S, Yamauchi KI, Torii Y, Kawaski M, Sakano S Clinical radiographic and thermographic assessment of osteoarthritis in the knee joints. *Annals of the Rheumatic Diseases* 2002, 61(9): 852–854.

6 Potential of Thermal Imaging to Detect Complications in Diabetes

Rationale for Diabetes Screening with Thermal Imaging

6.1 INTRODUCTION

Diabetes mellitus (DM) is a complex, chronic, and metabolic disorder characterized by high blood glucose levels also termed hyperglycemia (Zimmet et al., 2014). Insulin is an essential hormone produced by the beta cells in pancreas which is involved in the regulation of the amount of glucose circulating in the blood. The inability to produce insulin secretion or insulin resistance leads to hyperglycemia which is the clinical indicator of DM (Abdelalim, 2020). Diabetes is a life-threatening health disease which is etiologically classified into three types, namely Type I, Type II, and gestational DM.

Type I DM is also known as insulin-dependent diabetes or Juvenile diabetes, which is caused by an absolute deficiency of insulin production. This is attributed to the destruction of insulin-producing β-cells by the auto-immune system of the host. Type I DM mechanism is illustrated in Figure 6.1.

The loss of insulin production leads to improper regulation of the blood glucose levels due to disturbed transport of glucose molecules into cells. The etiology of type I DM is still largely unexplored and ways to prevent it are still a topic of wider research (Rother, 2007). The symptoms of Type I disease are excessive thirst, frequent urination, blurred vision, lack of energy, fatigue, and unexplained loss of weight (Facchinetti et al., 2014). Type II DM is also known as non-insulin dependent diabetes. It's pathomechanisms are explained by both a relative deficiency of insulin and a resistance of target tissues such as liver, muscle, and fat cells against insulin. A simple description of insulin resistance is the diminished effect of a unit of insulin on the quantum of lowering blood glucose. The type II

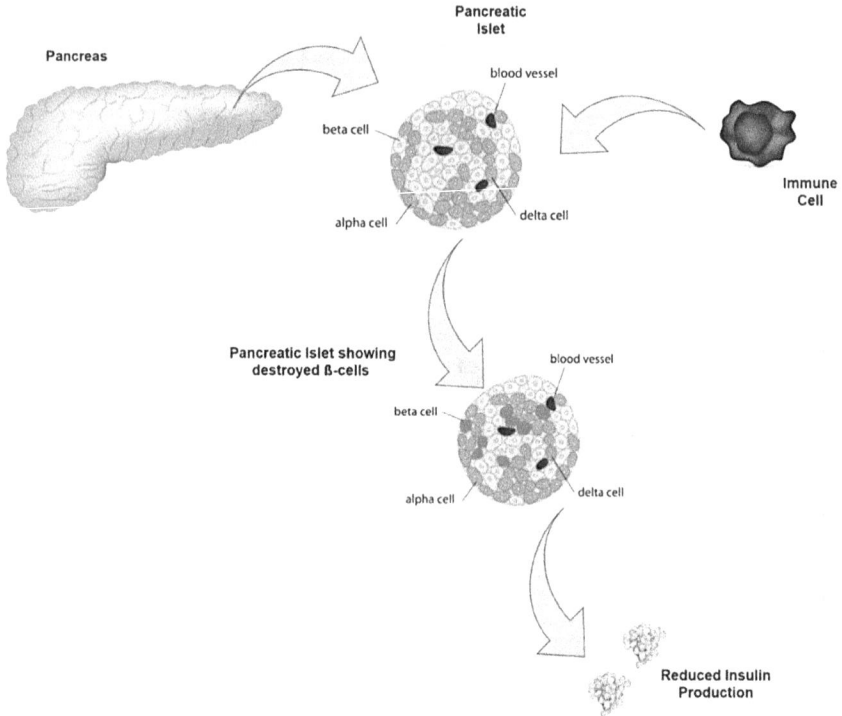

FIGURE 6.1 Type-I diabetic condition mechanism due to reduced insulin secretions because of destruction of ß-cells by the host autoimmune cells.

DM mechanism is illustrated in Figure 6.2. Globally, in around 90% of the total population, the Type II DM is most common, and its development is associated with sedentary lifestyle, poor diet, lack of exercise, and obesity (Zimmet et al., 2014). The occurrence of Type II DM could be prevented, or the onset of the disease could be hindered by the lifestyle alteration and administration of oral drugs (Rother, 2007). The symptoms are similar to type I DM but in some cases, they are found to be symptomless and it is quite difficult to determine the onset of the disease (Thirunavukkarasu, Umapathy, Krishnan et al., 2020b). Insulin replacement and cell therapy are widely researched for the treatment of these two types of diabetes disease.

6.1.1 Diagnosis

Criteria to diagnose diabetes mellitus are internationally well established. Diagnosis is based on fasting plasma glucose value (FPG) and plasma glucose obtained after an oral glucose tolerance test (OGTT). In the latter test, the 2-h postload glucose (2-h PG) is used for discrimination. Pre-diabetes is a metabolic condition situated between normoglycemia and diabetic hyperglycemia. The

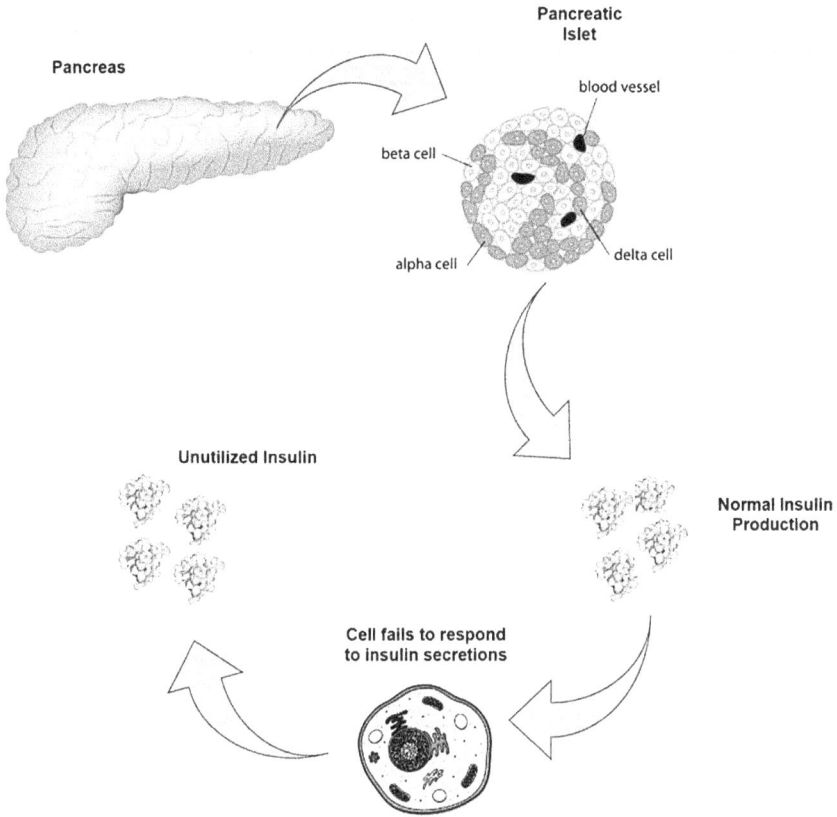

FIGURE 6.2 Type-II diabetic condition mechanism which is due to insulin resistance of the cells.

pre-diabetic phase can also be identified by using FPG and OGTT. FPG is the recommended test to screen for diabetes. Table 6.1 provides cut-off plasma glucose values for normoglycaemia, pre-diabetes, and hyperglycemia.

Although blood sampling via venipuncture is uncomfortable, sometimes painful, and, in rare cases, associated with infections, no other sufficiently reliable method is available to determine the level of plasma glucose.

In both Type-I and Type-II diabetic subjects, the reduced insulin in the blood impacts the metabolic rate of glucose molecules (Expert Committee on the Diagnosis and Classification of Diabetes Mellitus, 2003). Thermal imaging or IR imaging system allows the interpretation of the human body surface temperature distribution with minimal discomfort to the patient. The recorded temperature variation depends on the composite associations defining the heat exchange processes between skin tissue, superficial vasculature, and metabolic activity. The modality measures metabolic rate in terms of skin temperature variations which are not effortlessly measured by any other methods. Thermal imaging is non-invasive, as it

TABLE 6.1

Glucose Level for Normal, Pre-diabetic, and Diabetic Subjects

Normoglycaemia	Pre-diabetes	Hyperglycemia (diabetes[*])
FPG < 110 mg/dL	FPG ≥ 110 and > 126 mg/dL	FPG ≥ 126 mg/dL
2-h PG < 140 mg/dL	2-h PG ≥ 140 and > 200 mg/dL	2-h PG ≥ 200 mg/dL
		Symptoms of diabetes and casual plasma glucose concentration ≥ 200 mg/dL

[*] A diagnosis of diabetes must be confirmed, on a subsequent day, by the measurement of FPG, 2-h PG, or random plasma glucose (if symptoms are present).

inertly captures the radiant heat emitted from the human body. So, it is free of ionizing radiations, relatively inexpensive, and can be readily used as an assistive tool to the other standard imaging methods.

6.1.2 THERMOGRAPHY IN DIABETES EVALUATION

The advent of new technologies in recent years measures the skin surface temperature to assist in the diagnosis of diseases. A deviation in core body temperature or asymmetric skin temperature distribution is a hallmark of improper functioning of the human body and indicates the possibility of a disease (Ring and Ammer, 2012). Infrared thermography provides thermal information that may assist the interpretation of the physiological condition of the human body (Francis Ring, 2012).

Early works on the application of thermography in the study of diabetes mellitus could be attributed to the works of Brånemark et al. (1967) who studied the thermal emission patterns of 16 diabetic subjects in 1967. The study revealed that the thermal patterns of all the diabetic subjects showed abnormalities in both the feet and hands. However, the emission patterns were significantly pronounced in feet compared to the hands.

The scientist Francine J. Prokoski, in the mid-1960s, reported that facial thermograms are found to be highly unique to individuals and a possible method for identification, although even identical twins may have different facial thermograms. Facial infrared thermography captures the thermal patterns generated partly by the facial blood vessels. Facial thermal images may indicate an underlying disease. It was also reported that facial infrared thermography may serve as an alternative to the conventional clinical screening method for diabetes (Thirunavukkarasu, Umapathy, Janardhanan et al., 2020a). However, the thermal patterns captured at the facial surface are influenced by many factors such as gross anatomy of the subject's head, vascular or circulatory system of the person, cosmetic applications on the surface of the face, immediate contact with ice pack, cigarette, or intake of hot beverages, recent physical activity, and also by ambient temperature (Ammer and

Ring, 2019). So, stringent measurement protocol is to be followed for any kind of reliable insights from the facial thermograms.

Diabetic foot (DF) is one of the severe difficulties faced by diabetic patients. An early detection and appropriate management of diabetic foot problems can prevent distressing consequences such as limb amputation. Several researches have confirmed that temperature variations in the plantar region can be related to diabetic foot problems. Thermal imaging has been successfully validated to detect impediments related to diabetic foot, as it is offered as a fast, non-contact, and non-invasive technique to visualize the temperature variations of the feet.

This chapter discusses the rationale for using thermal imaging for screening diabetes conditions and their complications through the following objectives:

(i) To study the various anthropometrical variables and study the thermal patterns (°C) in the facial region such as forehead, eyes, nose, cheek, chin, tongue and also in foot regions between the healthy and diabetic subjects.

(ii) To examine the implications of using thermography to identify and detect complications of diabetes such as diabetic foot and foot ulcers.

(iii) To review the applications of machine-learning- and deep-learning-based approaches for the screening of diabetes and its complications such as the diabetic foot ulcers based on thermograms.

6.2   SELECTION OF STUDY POPULATION

The selection of volunteers for the study of diabetes screening from body thermography requires strict scrutiny of the test subjects by means of questionnaires and also other conditions such as the inclusion and exclusion criteria.

The criteria provided in Figure 6.3 are obtained from a study approved by the ethical committee of SRM Hospital and Research Centre, Kattankulathur, Tamil Nadu, India (Ref. No.: 834/IEC/2015). The investigators of the original work organized a medical camp and after obtaining the required consent from the volunteers, facial and tongue images of the test subjects were obtained (Thirunavukkarasu, Umapathy, Janardhanan et al., 2020a). Moreover, some of the anthropometrical variables such as

- weight (kg),
- height (cm),
- Blood pressure (DBP) (mmHg),
- waist circumference (WC) (cm), and
- hip circumference (HC) (cm)

were measured from the volunteers. For the blood glucose analysis, a 5 mL blood sample was drawn to obtain fasting blood glucose (FBG (mg/dL)), post-prandial blood glucose (PPBG (mg/dL)), and glycated hemoglobin content (HbA1c (%)) using standard procedure.

INCLUSION CRITERIA

- Participants aged from 20 to 50 years
- Healthy and Type II Diabetic volunteers with overnight fasting condition (8 hours)

EXCLUSION CRITERIA

- Hypo and hyperthyroidism
- Cardiovascular diseases
- Fever
- Anaemia
- Pregnant and nursing women

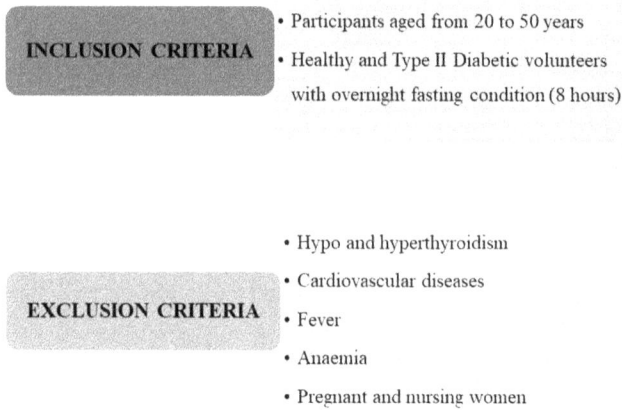

FIGURE 6.3   A typical study population of diabetic test subjects with the inclusion and exclusion criteria.

Similarly, for assessing the diabetic complications such as diabetic foot, proper inclusion and exclusion criteria is to be followed. Volunteers are generally advised to stop medication before the thermal image acquisition. Subjects who do not exhibit any signs of any sickness at the acquisition time of the thermograms are selected for the study. Moreover, volunteers with an underlying medical condition of amputation, prior episode of foot ulcers, foot infections and fractures, surgery of lower limbs, and other acute peripheral vascular disease shall be excluded. In addition, subjects with substance and alcohol abuse shall also be excluded.

6.2.1   SUBJECT PREPARATION

Thermal analysis of body regions such as face, abdomen, and foot regions in relation to blood glucose requires careful experimental setup. Following standardized protocols for thermal imaging such as the guidelines provided by the Glamorgan protocol developed at the University of Glamorgan (Ammer 2008) and the International Academy of Clinical Thermology (IACT) "International Academy of Clinical Thermology Quality Assurance Guidelines Standards and Protocols in Clinical Thermographic Imaging" 2020), the test subjects and the environment conditions shall be monitored in the research study. Following the protocol reduces the impact of ambient environment on the test subjects, thus leading to proper thermal analysis. The volunteers shall be provided necessary information and guidelines such as the following shall be followed: (a) duration of 8 h of fasting for measuring FBG, (b) duration of 2 h after prandial for PPBG, (c) no consumption of liquor, caffeine, or energy drinks, (d) no application of skin creams or oil on the body regions that are to be imaged, (e) no exercising prior to thermal imaging, (f) no chewing of tobacco, gums, or candies, (g) no smoking prior to imaging, and (h) no medication during the day of the thermal imaging. In addition, the room where the subject thermograms are obtained shall be maintained at a constant temperature of 22–25°C. The subjects were provided a 15 min time to acclimate to the measuring environment and were asked to be in a resting or supine position based on the region of interest (ROI). A thermal imaging camera with a minimum resolution of (320×240) is then used to acquire the regional thermograms with a distance in the range of 1–2 m based on the anatomical region.

6.3   MEASUREMENT PROTOCOL

The room utilized for thermal imaging must meet some basic prerequisites and should be of an adequate size to keep up the homogeneous temperature. Also, there must be satisfactory space for positioning the thermal camera and participants of all sizes to acquire their thermal images. To meet these necessities, a room size of 2×3 m or 3×4 m is preferable for thermal imaging. At lower temperatures, the participants are likely to be shivering and at higher temperatures, the participants may feel sweating. So, the room temperature should be in such a way that the participant's physiology is not adjusted to the point of shivering or sweating (Ammer 2008).

The ideal ambient room temperature is in the range from 18°C to 25°C but should be maintained within ± 0.5°C/h because it is the most important requirement for thermal imaging. The temperature of the imaging room is monitored continuously by a room thermometer so that it is maintained constant. The imaging room humidity should be controlled so that there is no air moisture developing on the skin that can interact with infrared thermal radiation. The windows and doors of the imaging room should be completely sealed and incandescent lights should not be used while imaging so that naturally emitted infrared radiation from the human skin can be captured by the camera (Ring and Ammer, 2012; Vollmer and Möllmann, n.d.). The above-mentioned procedures shall be followed during the clinical trial process for reliable data acquisition. The schematic representation of the experimental setup for infrared thermal imaging is depicted in Figure 6.4 which is used for diabetes screening.

6.4   STUDY OF THERMOGRAMS IN SUBJECTS WITH DIABETES

The preliminary analysis of evaluating the temperature variations in different anatomical regions between the two classes, namely normal and diabetic groups published in literature (Thirunavukkarasu, Umapathy, Janardhanan et al., 2020a) is discussed here. In this section, we review the thermal analysis of facial thermograms with respect to the two classes.

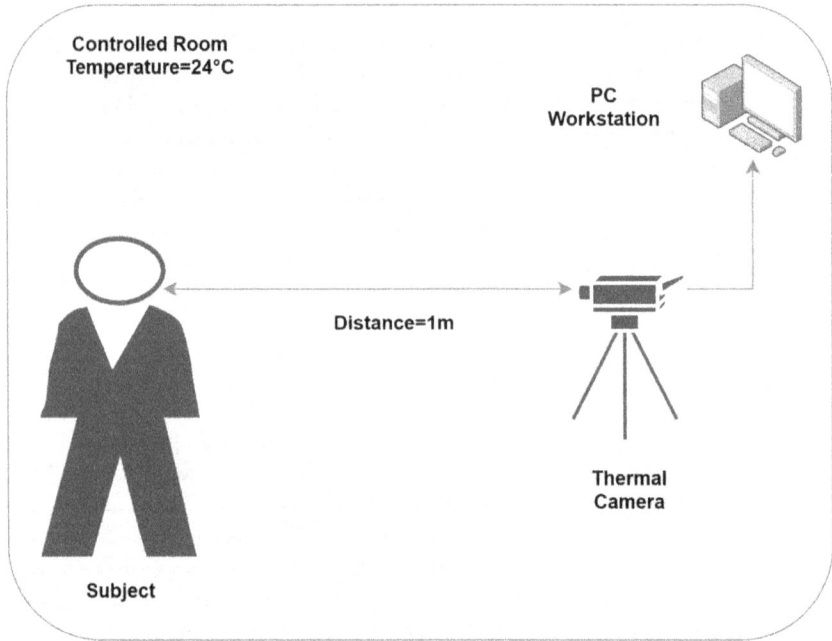

FIGURE 6.4   The experimental setup of acquiring regional thermograms from test subjects.

6.4.1   FACIAL THERMAL IMAGING MEASUREMENTS

Facial thermograms which are obtained from the normal and diabetic groups were analyzed to find any deviations in the temperature profile. So, some ROIs were selected in the facial region to determine the temperature. For the research study, three ROIs were selected, namely T_{ROI}, T_{MAX}, and T_{TOT}. Figure 6.5 illustrates in detail the templates of the three evaluation methods (Thirunavukkarasu, Umapathy, Janardhanan et al., 2020a). The temperature values were measured by using the measurement tools manually provided with the FLIR Tools software.

- T_{ROI}: This ROI is defined by the prominent facial markers such as the forehead, eyes, nose, and chin. The mean temperature values are measured by placing the measurement tool on the specific ROI.
- T_{MAX}: This ROI is defined by the regions of maximal temperature present in the face regions, namely the eye corners and over the mouth.
- T_{TOT}: This is the aggregate value for the whole face region.

6.4.1.1   Temperature Measurement Analysis

Facial thermograms are obtained from the two subject groups and studied to determine the temperature profile based on the three ROIs, namely T_{ROI}, T_{MAX}, and T_{TOT}. First, the ages of the two groups are matched and blood glucose parameters are obtained, which is reported in Table 6.2. FBG shows an average value of

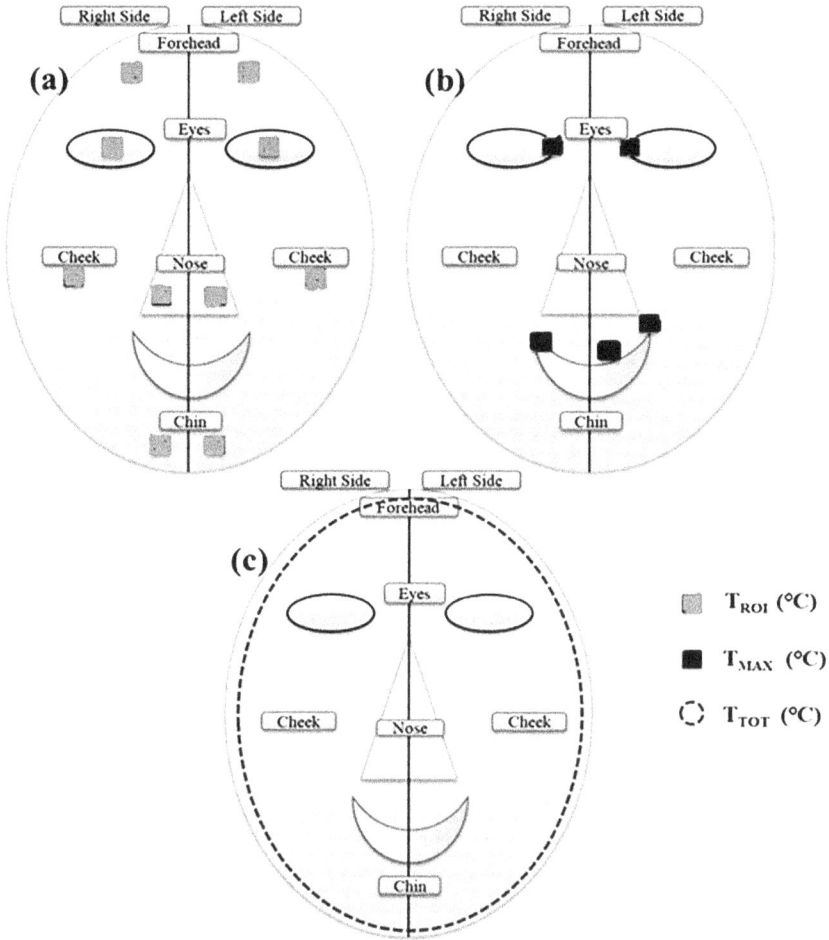

FIGURE 6.5   Templates of three statistical methods: (a) T_{ROI}, (b) T_{MAX}, and (c) T_{TOT}.

87.3 ± 18.5mg/dL and 148.7 ± 77.9mg/dL for healthy and diabetic groups, respectively. There is a significant difference ($p < 0.05$) in the blood glucose concentration during fasting between the two groups by using Student's t-test. PPBG also shows a significant ($p < 0.05$) increase in the PPBG level signifying low blood glucose control in the diabetic subjects. This is also reflected in the HbA1c values showing the least glucose control in the diabetic subjects. When performing the thermal analysis of the two groups based on the selected ROIs, we also observe that some of the regions show significant differences in the temperature profile of the two groups.

T_{ROI} Forehead shows an average value of 35.5 ± 0.5 °C and 35.3 ± 0.5 °C for the healthy and diabetic groups signifying that there is a drop in the temperature around this region for diabetic subjects. The T_{ROI} Eyes regions also show a

TABLE 6.2

Parameters of the Study Population (N = 160)

Parameters	Group I Healthy Subjects (N = 80)	GROUP II Diabetic Subjects (N = 80)	p-Value S (p < 0.05) NS (p > 0.05)
Age	41.2 ± 10.8 years	42.9 ± 9.6 years	NS
FBG	87.3 ± 18.4 mg/dL	148.7 ± 77.9 mg/dL	S
PPBG	114.7 ± 15.2 mg/dL	228.45 ± 100.1 mg/dL	S
HbA1c	5.6 ± 0.4%	8.6 ± 1.7%	S
T_{ROI} Forehead	35.5 ± 0.5°C	35.3 ± 0.4°C	S
T_{ROI} Eyes	35.2 ± 0.4°C	34.9 ± 0.4°C	S
T_{ROI} Nose	33.4 ± 1.6°C	32.6 ± 1.2°C	S
T_{ROI} Cheek	34.0 ± 0.7°C	33.9 ± 0.7°C	NS
T_{ROI} Chin	35.1 ± 0.5°C	35.0 ± 0.6°C	NS
T_{MAX}	36.9 ± 0.3°C	36.6 ± 0.4°C	S
T_{TOT}	34.8 ± 0.5°C	34.6 ± 0.5°C	S

Source: Thirunavukkarasu, Umapathy, Janardhanan et al. (2020a)
(S) – significant, (NS) – non-significant
FBG – fasting blood glucose, PPBG – post-prandial blood glucose, HbA1c – hemoglobin A1c, T_{ROI} – temperature in the region of interest, T_{MAX} – temperature maximal, and T_{TOT} – total temperature

significant drop in the temperature for the diabetic subjects. Similarly, the temperature of the T_{ROI} Nose regions provides significantly lower values for the diabetic groups. However, temperatures in the chin and cheek regions did not provide a reasonable difference in the temperature for the two subject groups. However, the temperature at the maximal points such as the corner of the eyes provided significantly lower values for the diabetic groups. When determining the temperature considering the whole face, T_{TOT} also provided lower temperature values for the diabetic groups which are significantly different from that of the healthy individuals. The thermograms of the test subjects for the two groups are illustrated in Figures 6.6 and 6.7 for the normal and the diabetic groups, respectively. The images are provided with measurement tools provided in the thermal analysis software for handy temperature measurement reading. Comparing both the thermograms, we observe that temperatures of the diabetic test subject show predominantly lower temperature readings in the selected ROIs which is clearly visible in the grayscale color spectrum of the thermograms. Hotter areas are provided in white color and cooler areas are provided in gray colors.

The temperature variations between the two subject groups provide the necessary insights that thermograms could be used to detect diabetes in the affected individuals based on thermal signatures in the facial regions. Facial thermography could be a useful screening device in detecting diabetes conditions when following proactive measurement procedures and stringent inclusion and exclusion

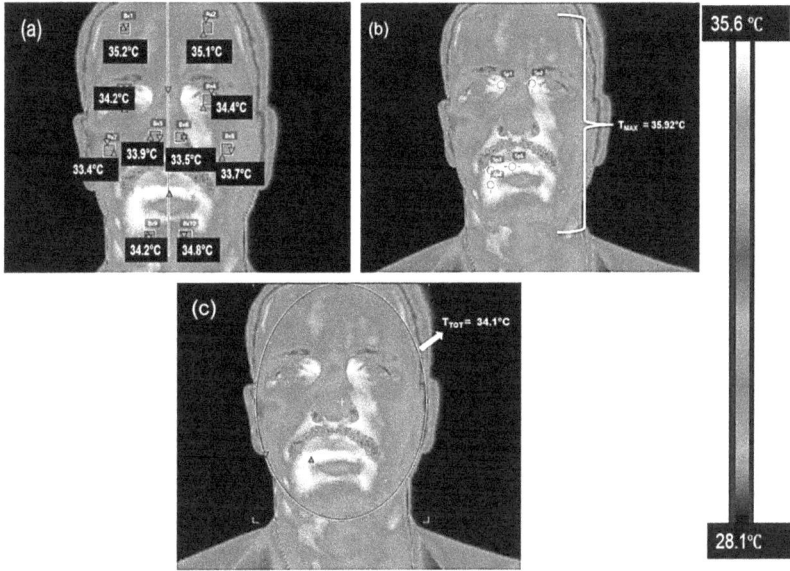

FIGURE 6.6   Measured facial skin surface temperature of a normal subject: (a) measured average temperature in the region of interest (T_{ROI}), (b) maximal temperature (T_{MAX}), and (c) total temperature (T_{TOT}). (See Thirunavukkarasu, Umapathy, Janardhanan et al., 2020a.)

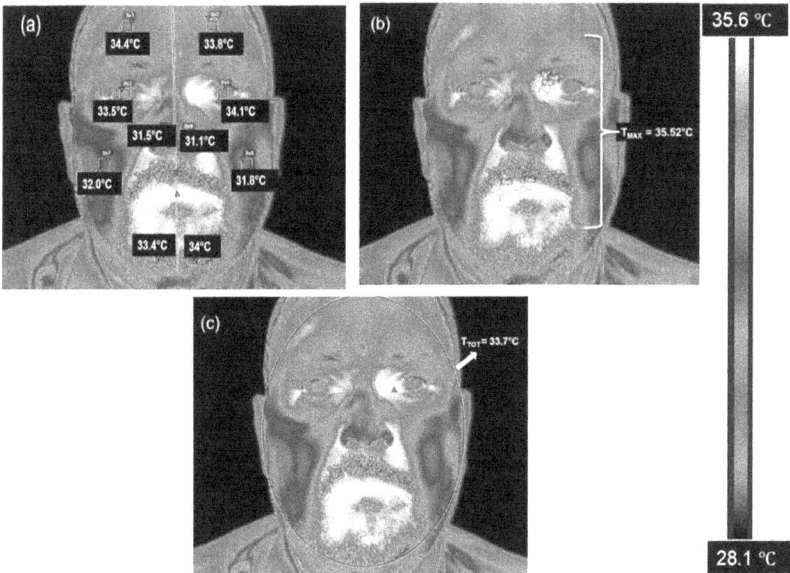

FIGURE 6.7   Measured facial skin surface temperature of a diabetic subject: (a) measured average temperature in the region of interest (T_{ROI}), (b) maximal temperature (T_{MAX}), and (c) total temperature (T_{TOT}). (See Thirunavukkarasu, Umapathy, Janardhanan et al., 2020a.)

criteria for the study population. However, the usability of facial thermal imaging in diabetic screening applications for the general population is quite limited since heat is a by-product of metabolism and does not have a specific role in physiological function and also requires much more investigation.

6.4.2 Thermal Imaging in Diabetic Foot

DF ulcer is a prevalent complication of diabetes disease that leads to hospitalization and medical care. DF ulcer is often caused by uncontrolled high levels of glucose concentration in the body that leads to the damage of blood vessels, especially in the foot regions. This condition leads to a state of pain or numbness in the foot regions thus leading to peripheral neuropathy. Patients unaware of the foot problems could lead to stress in the feet which might cause callus, burns, or cuts. These tissue conditions if not treated in time, could lead to foot ulcers. Severe conditions of foot ulcers require treatments such as amputation of the parts of the foot or the lower limb. The rate at which amputations are performed in the diabetic population is typically 10–20 times the general population. The incidence of a diabetic foot condition that leads to amputation is around 1.5–3.5 per 1000 persons per year. So, proper management of diabetic foot is necessary for better patient care.

Diabetic foot ulcer is characterized first by the formation of inflammation in the affected regions which is caused by damaging of the blood vessels due to high levels of glucose in the circulating blood. It then leads to neuropathy in the foot regions which, when left unnoticed, develops into foot ulcers. In addition to the neuropathy, other factors also increase the risk of diabetic foot ulcers, namely (i) insufficient blood supply to the foot regions, (ii) obesity, (iii) improper footwear that leads to skin damage and breakdown, and (iv) poor healing of the wound due to peripheral arterial disease which is often found in diabetic subjects (Lawrrence A. Lavery and Armstrong, 2007). Recently, the application of thermal imaging in studying the temperature variations of severe diabetic subjects in the foot regions due to inflammation has captured the interest of researchers. This is due to the advancement of thermal sensors which now provides improved thermal resolution and temperature sensitivity. To illustrate the application of thermal imagery in diabetic foot analysis, we provide here a few sample foot thermograms of normal and diabetic subjects based on the work (Hernandez-Contreras et al., 2019). Figure 6.8a shows the thermal imaging of subjects with a severe diabetic condition. The temperature pattern in the left foot is different compared to the contralateral region of the other. This is a common feature signifying peripheral neuropathy. However, for the normal subjects shown in Figure 6.8b, we could observe that the temperature patterns in both the legs are similar.

One of the earliest works involved in the study of diabetic foot abnormality using thermal imaging could be found in the works of Ammer et al. (2001). The seminal work studied a cohort of 76 subjects with a prehistory of Type-II diabetes and found that around half of the subjects showed increased plantar temperature. However, it was reported that there was no strong correlation to the skin changes

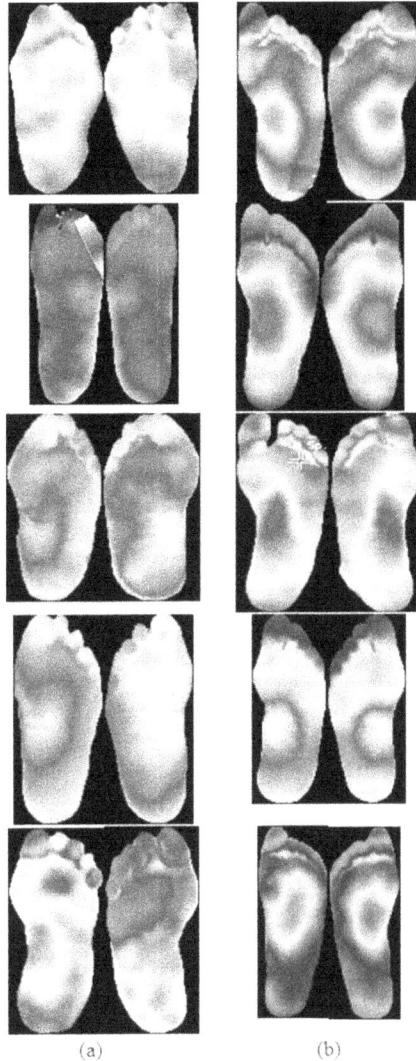

FIGURE 6.8 Thermal imaging of foot regions: (a) diabetic subject and (b) normal subject. (See Hernandez-Contreras et al., 2019.)

and hot spots which were identified in the foot regions. The authors concluded that elevated temperature in the foot regions could be used as a predictive marker for the formation of ulcers.

The temperature variations in the foot regions are more pronounced in diabetic subjects compared to the control group which might result in ulcerations. There is elevated temperature found not only in the affected regions but also observed in the entire foot region (Harding et al., 2002). The presentation of temperature variations in diabetic foot subjects is also studied by researchers for ulcer detection based on asymmetry analysis. Kaabouch et al. developed a methodology to compare the temperature profile of the two foot regions and performed an intensity-level processing to determine the disparity of the pixel intensity values between the identical regions of the feet. A threshold scheme was proposed to reduce the false detection of inflammation (Kaabouch et al., 2009). However, the approach assumed the size and shape of the left foot and the right foot to be uniform for inflammation detection, which could not be applicable for the general population.

Lavery et al. conducted an extensive randomized trial to study the use of thermal imaging for the management of diabetic foot ulcers at home. Their initial study comprised 85 diabetic foot risk subjects who were given two sets of therapy. The first set of subjects involved standard therapy which involved the usage of diabetic footwear and regular evaluation by a physician. The second set of participants was given enhanced therapy based on regular monitoring of the foot regions using thermography. They concluded that enhanced therapy provided a better way of preventing the occurrence of ulceration in the foot (Lavery et al., 2004). In a similar study by Lavery et al., the authors performed another randomized trial comprising 173 subjects and studied the efficacy of using thermal imaging of the foot regions for management of diabetic foot ulcer formation. Their study concluded that infrared imaging at home could be used as an effective early warning system for foot ulcers (Lavery and Armstrong, 2007). They reported that only 8.5% of volunteers had foot ulcers which had regular home monitoring using thermography compared to other standard therapies. Their seminal work formed the basis of using the elevated temperatures (greater than 4°F (2.2°C) compared with the contralateral foot) as a marker for being "at risk" of ulceration formation due to inflammation at the site of measurement.

A further enhancement in the early detection of foot ulcers could be found in the works of Netten et al. where the group developed an experimental setup for capturing foot thermograms and obtaining the mean difference of the foot regions in comparison to the contralateral regions. They conducted their study using 15 diabetic subjects with different degrees of complications in the foot regions. Their study reported that there were no complications in the diabetic groups when the mean temperature was less than 1.5°C between the ipsilateral and contralateral foot regions. However, there was a difference of more than 2°C for patients with local foot complications and the difference is even more than 3°C for severe foot complications. Their work suggested that thermography could be a promising tool in foot care management that has the ability to detect early signs of diabetic foot complications (van Netten et al., 2013).

Detection of foot complications due to diabetes based on thermal imaging required a fair amount of manual annotation to segment the infected regions or hot regions for proper analysis. With the advancement of dedicated hardware for running machine learning algorithms, it was plausible to use for the management of diabetic foot ulcers. Liu et al. (2015) developed a machine-learning-based system for the automated detection of diabetic foot infections through the infrared imaging tool. Their work involved automating the asymmetry analysis of temperature profiles in the contralateral foot regions. The measurement points on the left foot and right foot were initially registered and then the temperature differences were calculated. They reported a sensitivity of 97.8% and specificity of 98.4% when studied on 76 high-risk diabetic patients with manual annotation as a reference.

Adam et al. (2018) proposed a computer-aided diagnosis system for the detection of diabetic foot ulcers with higher-order statistics (HOS) image features. The work involved transforming the input RGB thermal images into a wavelet domain and the HOS features were extracted. Then, some of the state-of-the-art machine learning classifiers such as support vector machine (SVM) were trained on the extracted features for the classification of diabetic foot ulcers. The authors reported a maximum accuracy of 89.39% and suggested that their automated system could be used as a second opinion tool for diabetes foot management.

Diabetes condition in individuals disturbs the thermoregulation of the human body. The relation to obesity condition requires clinical evidence for a better understanding of the effects of diabetes. Renero studied the complex relation of thermoregulation to obesity condition from plantar skin thermograms for diabetic foot assessment (Renero 2017). Their work obtained plantar thermal images of healthy, obese, and diabetic subjects. They reported interesting results suggesting that the cold part of the plantar skin region was more prone to the formation of infections or diabetic foot than the hot regions on the asymmetric thermograms (Renero 2017).

More recently, IRT was used in conjunction with standard methods of assessing foot, such as the toe pressure measurement for the diagnosis of vascular disorders in diabetic subjects. Ilo et al. conducted a clinical trial of 118 DM subjects and 93 control subjects to assess the application of thermal imaging in foot trauma. Their work reported that mean temperature was highest in DM subjects with neuroischemia. Moreover, the hot regions were presented with high toe pressure, thus suggesting that IRT with other tools such as toe pressure measurement could provide useful screening of vascular disorders in feet (Ilo, Romsi, and Mäkelä, 2020).

Furthermore, Hernandez-Contreras et al. created a comprehensive plantar thermograms database which is available in public domain for aiding researchers in the development of automated detection of diabetic foot problems based on intelligent systems. The dataset contains 334 individual plantar thermograms from 122 DM and 45 control subjects. The images are pre-processed such that the foot is isolated from the background radiation and manually corrected to work directly with it (D. A. Hernandez-Contreras et al., 2019, D. A. Hernandez-Contreras et al., 2017). Figure 6.8 shows the sample thermal images of plantar regions provided in the dataset. Table 6.3 shows an excerpt of temperature information provided in the original dataset of the control group for different angiosome zones. Table 6.4 shows an excerpt of temperature information provided in the original dataset of the diabetic group for different angiosome zones. Exploratory data analysis of this database is provided here for more information to understand the correlations

TABLE 6.3
Patient Information with Plantar Skin Temperature of the Control Group (Sample of Healthy Subjects)

Subject	Gender	Age (years)	Weight (kg)	Height (m)	Right Foot						Left Foot					
					General (°C)	LCA (°C)	LPA (°C)	MCA (°C)	MPA (°C)	TCI	General (°C)	LCA (°C)	LPA (°C)	MCA (°C)	MPA (°C)	TCI
CG001	M	25	67	1.8	25.9	26.0	25.4	26.3	25.8	0.1	25.4	25.8	24.8	26.0	25.3	0.4
CG002	M	26	80	1.6	28.0	27.6	27.7	28.0	28.4	1.9	28.0	27.7	27.7	28.5	28.3	2.0
CG003	M	24	60	1.6	29.5	30.0	29.1	30.2	29.7	3.7	29.4	30.1	28.9	30.1	29.0	3.5
CG004	F	22	65	1.5	27.8	27.6	27.5	28.5	27.5	1.8	27.6	27.5	27.3	28.2	26.8	1.5
CG005	F	38	60	1.5	26.2	26.3	25.6	26.8	26.2	0.2	26.7	26.8	26.1	27.3	26.8	0.8
CG006	F	52	69	1.5	26.5	27.3	25.7	27.7	26.2	0.7	26.8	27.5	26.1	27.9	26.5	1.0
CG007	F	25	60	1.5	26.1	25.9	25.6	26.6	26.2	0.2	26.4	26.4	26.1	26.0	26.9	0.5
CG008	F	30	57	1.5	24.3	23.1	24.7	23.5	25.0	1.8	24.6	24.0	24.9	23.5	24.3	1.7
CG009	M	30	70	1.6	24.7	23.7	24.3	24.8	25.5	1.3	24.9	24.2	24.5	24.9	25.7	1.1
CG010	M	26	63	1.8	26.9	27.2	26.3	28.1	26.8	1.1	26.8	26.7	26.4	27.5	26.6	0.8
CG011	F	37	63	1.5	26.2	25.8	26.1	26.2	26.4	0.3	26.0	25.8	25.7	26.1	26.0	0.1
CG012	F	25	70	1.6	27.9	29.0	27.1	29.0	27.6	2.2	27.6	28.5	26.5	29.4	27.7	2.0
CG013	M	25	63	1.6	26.3	26.4	25.7	26.7	26.4	0.3	26.7	27.2	26.1	27.3	26.2	0.7

Source: Medial Plantar Artery (MPA), Lateral Plantar Artery (LPA), Medial Calcaneal Artery (MCA), and Lateral Calcaneal Artery (LCA). TCI Is the Thermal Change Index Based on Asymmetry Analysis Developed in the Works of D. Hernandez-Contreras et al. (2019)

TABLE 6.4

Patient Information with Plantar Skin Temperature of the Diabetic Group (Sample of Diabetic Subjects)

Subject	Gender	Age (years)	Weight (kg)	Height (m)	Right Foot						Left Foot					
					General (°C)	LCA (°C)	LPA (°C)	MCA (°C)	MPA (°C)	TCI	General (°C)	LCA (°C)	LPA (°C)	MCA (°C)	MPA (°C)	TCI
DM001	M	55	82	1.66	32.4	32.0	32.7	32.3	32.7	6.4	32.0	31.5	32.3	31.8	32.3	6.0
DM002	M	60	90	1.71	26.9	26.5	26.9	26.8	27.3	0.9	26.3	26.3	25.9	26.7	26.4	0.3
DM023	F	50	63	1.48	29.1	26.8	30.4	27.5	30.1	2.7	30.4	28.6	31.4	29.3	31.3	4.1
DM024	M	60	72	1.66	28.7	27.5	29.4	27.8	28.9	2.4	32.4	31.1	33.2	31.3	32.8	6.1
DM025	M	63	80	1.66	32.0	30.8	32.8	30.9	32.5	5.8	32.1	30.5	32.9	31.0	33.0	5.8
DM026	M	56	104	1.68	32.3	31.8	32.6	32.1	33.0	6.4	32.3	32.5	32.2	32.5	32.2	6.3
DM027	M	51	74	1.60	32.4	31.8	32.9	31.8	32.4	6.2	31.9	31.4	32.3	31.2	32.2	5.8
DM028	M	57	79	1.68	25.2	24.7	25.1	25.1	25.7	0.8	24.1	24.1	24.0	24.1	24.4	1.8
DM029	F	70	76	1.50	27.8	27.3	27.9	27.8	28.2	1.8	29.0	28.7	28.9	29.7	29.6	3.2
DM030	F	46	59	1.45	28.4	27.7	28.6	28.0	29.5	2.4	28.2	27.8	28.2	28.4	28.7	2.3
DM031	F	76	54	1.46	30.5	30.8	30.1	31.3	29.7	4.5	31.7	30.8	32.2	31.4	31.0	5.4
DM032	M	45	93	1.71	32.3	30.0	33.8	29.8	34.1	5.9	31.1	29.4	32.1	29.3	32.5	4.8
DM033	M	57	56	1.68	25.0	25.0	24.7	24.6	25.4	1.0	23.9	23.2	23.9	23.4	25.2	2.0

Source: Medial Plantar Artery (MPA), Lateral Plantar Artery (LPA), Medial Calcaneal Artery (MCA), and Lateral Calcaneal Artery (LCA). TCI Is the Thermal Change Index Based on Asymmetry Analysis Developed in the Works of D. Hernandez-Contreras et al., (2019)

between different ROI of the foot regions between the two groups. Here, we calculate the Kendall correlation rank coefficient for the two subject groups. Kendall rank correlation coefficient (τ) measures the ordinal association between two variables. The value lies between -1 and $+1$, -1 indicating total negative correlation, 0 indicating no correlation, and 1 indicating total positive correlation. To calculate τ for two variables "X" and "Y," one determines the number of concordant and discordant pairs of observations. τ is given by the number of concordant pairs minus the discordant pairs divided by the total number of pairs.

The rank correlation coefficient of the two subject groups, namely the control group and the diabetic group is provided in Figures 6.9 and 6.10, respectively. Interpretation of the correlogram or correlation matrix allows us to analyze the relationship between each pair of numeric variables of the dataset. We could observe that there is a strong correlation between the contralateral regions of the angiosome in the control group compared to the diabetic group. This suggests that plantar skin temperatures show a non-uniform temperature distribution in DM which could be due to neuropathy or an early indication of diabetic ulcer. Thus, IRT could be a good prognostic tool in the detection of foot problems.

More recently, Vega et al. developed a deep-learning-based automated diabetic foot detection system which was trained on the plantar thermal image dataset

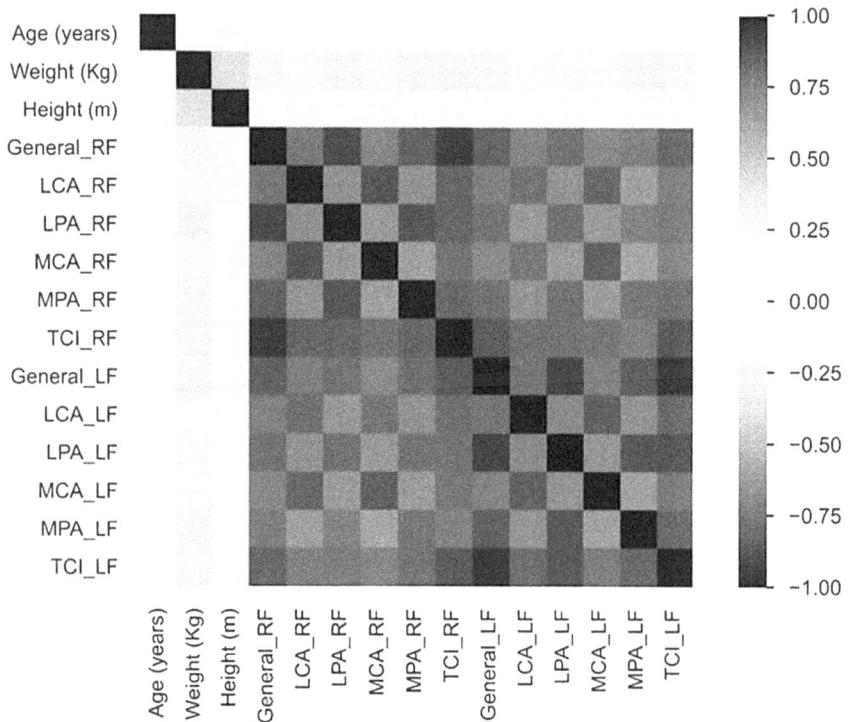

FIGURE 6.9 Correlation analysis (control group).

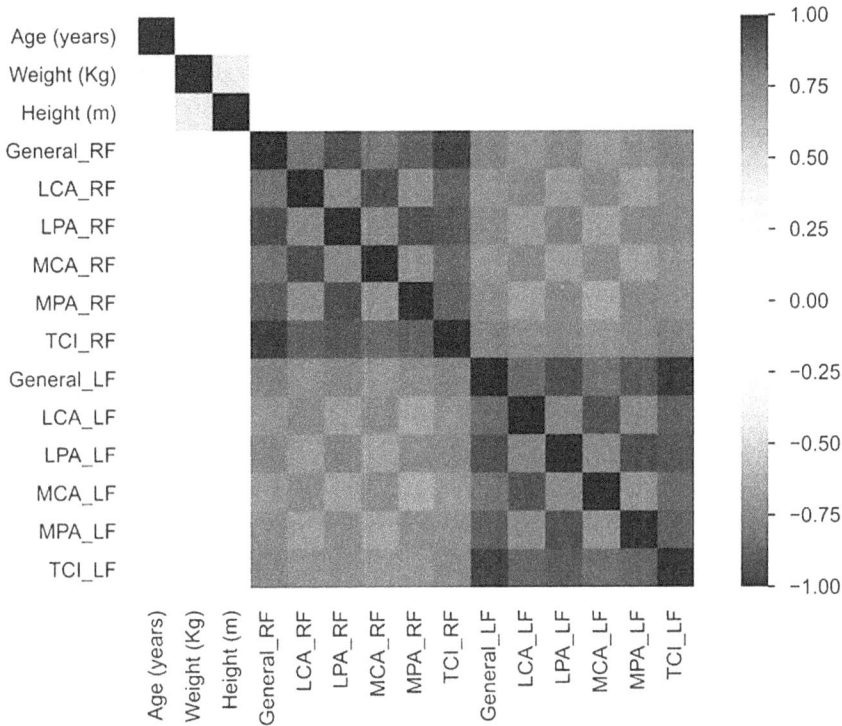

FIGURE 6.10   Correlation analysis (diabetic group).

described in the works of (D. A. Hernandez-Contreras et al., 2019). Their work compared machine learning-based techniques with Deep Learning (DL) structures. They have tested common structures based on the transfer learning approach, which involved AlexNet and GoogleNet. Moreover, they have designed a new DL structure, which was trained from scratch and was able to obtain higher classification values in terms of accuracy and other performance measures (Cruz-Vega et al., 2020).

6.5   COMPUTER-AIDED SYSTEM FOR SCREENING OF DIABETIC COMPLICATIONS

In this section, we review concepts that could be used in the development of automated diabetes complications screening system based on thermal images. The advancement of computer hardware which enables the implementation of state-of-the-art machine learning and deep learning algorithms for image classification problems has accelerated the applications in diabetes management. These systems, following a standard measurement protocol, could be a viable screen tool for the assessment of diabetes complications, namely the formation of diabetic foot in severe DM subjects.

6.5.1 FEATURES

In this section, we review a machine-learning-based diabetes complication screening system that could be used for automated thermal image classification. Feature-based image classification problems generally involve using feature extraction methods such as scale-invariant feature transform (SIFT), speeded-up robust features (SURF), oriented FAST (oFAST) and rotated BRIEF (rBRIEF) (collectively referred to as ORB), or morphological feature descriptors for image classification (Tareen and Saleem, 2018). The extracted features are used to train some of the state-of-the-art machine learning classifiers for the classification of medical conditions. Here, we review ORB-based feature extraction and classification for diabetes foot screening (Rublee et al., 2011).

6.5.1.1 Feature Extraction and Classification

Machine learning approaches rely on feature extraction processes for object detection, recognition, and classification. For computer vision problems and more specifically for image classification problems such as diabetes foot screening, there are a plethora of spatial and image features that are reported in literature. Some of the widely used image features are based on computing gray level co-occurrence matrix (GLCM) and extracting textural features such as contrast, correlation, moments, variance, entropy, homogeneity, and dissimilarity (Ashwin Kumaar and Thanaraj, 2015). Following that, a few researches also concentrated on extracting feature points or interest points for common computer vision applications such as image stitching, image matching, object detection, and recognition. One of the widely used and validated feature point extractor methods is the SIFT (Lowe, 2004b; Clemons, n.d.) and SURF (H. Bay et al., 2008). Though these methods provided excellent feature extraction performance, they are computation-intensive algorithms. A viable and alternate open-source feature extractor named ORB was proposed by Rublee et al. (2011), which provided good feature extraction performance with fewer computation subroutines. In this chapter, we review the application of ORB feature extractors for diabetes classification from anatomical thermograms.

6.5.1.2 Oriented Fast and Rotated Brief (ORB)

ORB is an efficient and alternative solution to some of the widely used feature point extractors and descriptors such as SIFT and SURF (Karami, Prasad, and Shehata, 2017; Tareen and Saleem, 2018). ORB is a combination of two key phases: (i) oFAST feature point extractor and (ii) rotated BRIEF feature descriptors. The procedure overcomes the important limitation of SIFT which requires good computation resources for determining the interest points in a given image. The following section explains the two steps of ORB feature extraction (Rublee et al., 2011).

a) **Oriented FAST Feature Point Selection**: FAST feature point extractor relies on determining the center pixel in a neighborhood radius of nine pixels and computing the intensity value. The feature point is selected if the pixel intensity is above the predefined threshold value. The number of points obtained by this process is more and hence the redundant and non-discriminant points are filtered by the Harris corner detection procedure. Based on the Harris corner, selected keypoints are ordered and the top-ranked "N" keypoints are finally selected. However, the FAST technique does not consider the orientation of the pixel in the selection of interest points. ORB enhances the FAST method by proposing oFAST which involves computation of intensity centroid that considers the orientation of the feature point selected. In addition, the aforementioned steps, they are also computed at different scales (multiscale pyramid approach) for proper selection of feature points.

b) **Rotated BRIEF Feature Descriptor**: After the selection of scale-invariant feature points based on the oFAST step, the points are vectorized by the BRIEF feature descriptors. It is a binary string of an image patch "p_c" obtained by running a set of intensity tests "t" on the selected patch of the image. It could be explained as follows:

$$t\left(p_c,\mathrm{x},\mathrm{y}\right)=\begin{cases}1 & : p_c\left(\mathrm{x}\right)<p_c\left(\mathrm{y}\right) \\ 0 & p_c\left(\mathrm{x}\right)\geq p_c\left(\mathrm{y}\right)\end{cases} \tag{6.1}$$

where $p_c(x)$ is the intensity of p_c at a point "x" and $p_c(y)$ is the intensity of p_c at a point "y." The image patches are smoothened by the Gaussian kernel around the selected feature point before the binary tests. Though the feature descriptors are simple to compute, the performance of the feature descriptors falls drastically when the feature point is rotated even slightly from the original position. Or, in a sense, the BRIEF features are rotational-variant. In order to make the BRIEF features rotational–invariant, similar to SIFT and SURF, a modification to the BRIEF features, namely rotated BRIEF (rBRIEF) is proposed in the original paper by Rublee et al. (2011). Here, the computation of BRIEF features is not computed in-plane directly, instead the BRIEF features are steered or rotated according to the orientation of the feature points for feature description. The experimental results in the seminal papers prove that the ORB provided superior performance in computation and accuracy of selection and description of interest points/feature points.

The anatomical thermograms from the two groups, namely the control group and the diabetic group are provided as input to the ORF feature extractor. Discriminant feature points are selected from the two groups and are used to train a set of machine learning classifiers such as Support Vector Machine (SVM), k-Nearest Neighbour (k-NN), Linear Discriminant Analysis (LDA), Random Forest (RF), and Gaussian Naïve Bayes (NB) for automated screening of medical conditions.

6.5.2   Automated Deep Learning Classifier

The classification of normal and diabetic foot thermograms could also be made fully automated through the application of the Convolution Neural Network (CNN). CNN provides optimized and automated feature extraction of the input images such as the plantar thermograms (Vardasca, Magalhaes, and Mendes, 2019). Applications of CNN in image classification are widely researched in computer vision field and are reported with state-out-the-art performance (Shamim Hossain, Muhammad, and Guizani, 2020). The availability of graphics processing unit (GPU) has provided immense computing power to accelerate the training and validation process of the deep learning architectures for image classification (Vardasca, Vaz, and Mendes, 2018).

6.5.2.1   Pre-trained CNN Classifier

In this section, we review a deep-learning-based approach for diabetes classification from anatomical thermograms by means of the transfer learning approach (Raghu et al., 2019; Alzubaidi et al., 2020). Here, CNN architecture which is trained on general images is used for diabetes classification. This is done by retaining the base structure and weights of the model (lower CNN layers) and tuning the higher layers of the deep learning network based on the input thermograms. Table 6.5 shows the modified CNN architecture that could be used for diabetes classification. The base model is a pre-trained VGG-16 model with 16 CNN layers (Simonyan and Zisserman, 2014). However, the top layers are modified to perform diabetes classification such as that of diabetic foot.

Figure 6.11 shows the modified CNN architecture used for diabetes classification based on anatomical thermograms. The lower level consists of CNN layers which are based on VGG-16 architecture and the top layers are modified to

---

TABLE 6.5

Pre-trained Model Based on VGG-16 for Diabetes Foot Classifications Based on Anatomical Thermograms

Layer (Type)	Output Shape	Parameters
VGG-16 (Base Model)	(1, 7, 10, 512)	14714688 (Pre-trained Weights)
(Flatten)	(1, 35840)	0
(Dropout)	(1, 35840)	0
(Dense)	(1, 1024)	36701184
(Dense)	(1, 512)	524800
(Dense)	(1, 64)	32832
(Dense)	(1, 32)	2080
(Dense)	(1, 2)	66

Total parameters: 51975650
Trainable parameters: 37260962
Non-trainable parameters: 14714688

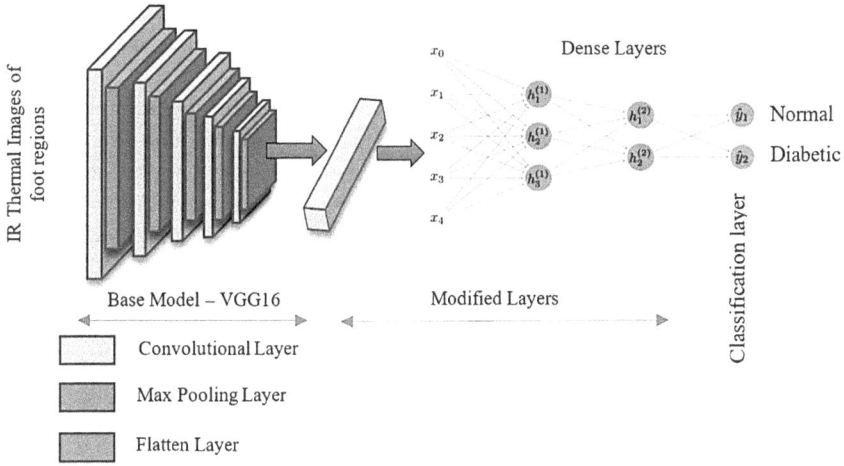

FIGURE 6.11 Modified pre-trained VGG-16 for diabetes complication screening from anatomical thermograms.

perform the transfer learning process for the diabetes classification. The top layers consist of fully connected layers which are the feedforward neural networks (F-NN) consisting of hidden nodes of 10245126432 and two output layers, respectively, for each layer. The top-most layer is the output layer which is the classification layer that uses Softmax activation. However, the other hidden layers are based on the ReLU activation function.

The transfer learning of the modified CNN model could be done through a dedicated GPU hardware. For example, the NVIDIA® Tesla® K80 Accelerator, which is provided as a free feature for the Google account users, uses Google Colab running TensorFlow deep-learning library. The input thermograms are usually resized to a size of 224 × 224 with three channels (RGB). Data augmentation methods such as zooming, scaling, shearing, and rotational, translational, horizontal, and vertical flipping could be used to increase the number of input thermograms for training deep learning models.

6.6 CONCLUSION

Diabetes mellitus is a long-term metabolic disorder which corresponds to a multitude of complications in the individuals. In this chapter, we showcased the novelty of using thermal imaging in applications of diabetes complications screening. First, the temperature analysis of facial thermograms of sex- and age-matched population groups showed temperature differences between the healthy and diabetic groups. However, relying only on the temperature information for diabetic condition is limited, since heat is a by-product of metabolism and does not have a specific role in physiological function. Prolonged uncontrolled high levels of glucose in the diabetic subjects may result in inflammation, trauma, ulcers, or severe infections such as osteomyelitis in the foot regions. The diabetic foot ulcer is a severe medical condition which could lead to disability and weakening of the quality of life in extreme cases.

The disease burden of DF diseases is expected to increase in the future due to the growing number of DM patients. So, an early detection of diabetic foot complications is vital for effective medical management. Various methods for thermal image analysis were presented. Among them, asymmetric temperature analysis is a widely used technique as it is straightforward to implement and provides satisfactory results. In addition, infrared-thermography-based CAD system for detecting diabetic foot is also discussed here. Also explained in this chapter are the necessary steps involved in using the thermograms to develop a computer-assisted diabetes complications screening system based on machine learning and deep learning approaches. This chapter provides necessary insights into screening of the diabetic foot syndrome which is a complication of diabetes where thermal imaging may have a role in screening and monitoring.

REFERENCES

Abdelalim EM Modeling different types of diabetes using human pluripotent stem cells. *Cellular and Molecular Life Sciences 2020 78:6* 2020, 78(6): 2459–2483. https://doi.org/10.1007/S00018-020-03710-9

Adam M, Ng EYK, Shu Lih O, Heng ML, Hagiwara Y, Tan JH, Tong JWK, Rajendra Acharya U Automated characterization of diabetic foot using nonlinear features extracted from thermograms. *Infrared Physics and Technology* 2018, 89: 325–37. https://doi.org/10.1016/j.infrared.2018.01.022

Alzubaidi L, Fadhel MA, Al-Shamma O, Zhang J, Santamaría J, Duan Y, Oleiwi SR Towards a better understanding of transfer learning for medical imaging: A case study. *Applied Sciences* 2020, 10(13): 4523. https://doi.org/10.3390/APP10134523

Ammer K The Glamorgan protocol for recording and evaluation of thermal images of the human body. *Thermology International* 2008, 18(4): 125–144.

Ammer K, Melnizky P, Rathkolb O, Ring EF. Thermal Imaging of Skin Changes on the Feet of Type II Diabetics. In *2001 Conference Proceedings of the 23rd Annual International Conference of the IEEE Engineering in Medicine and Biology Society*, 2001, 3:2870–2872. IEEE. https://doi.org/10.1109/IEMBS.2001.1017387

Ammer K, Ring F *The Thermal Human Body. The Thermal Human Body.* Jenny Stanford Publishing, 2019. https://doi.org/10.1201/9780429019982

Ashwin Kumaar M, Thanaraj P Feature extraction of arterio-venous malformation images using grey level co-occurrence matrix. *Indian Journal of Science and Technology* 2015, 8(35): 5. https://doi.org/10.17485/ijst/2015/v8i35/83387

Bay H, Ess A, Tuytelaars T, Van GL. Speeded-up robust features (SURF). *Comput Vis Image Underst* 2008, 110(3): 346–59. https://doi.org/10.1016/j.cviu.2007.09.014

Brånemark PI, Fagerberg SE, Langer L, Säve-Söderbergh J Infrared thermography in diabetes mellitus a preliminary study. *Diabetologia* 1967, 3(6): 529–32. https://doi.org/10.1007/BF01213572

Lowe D Distinctive image features from scale-invariant keypoints. *International Journal of Computer Vision*, 2004a, 60(2): 91–110.

Cruz-Vega I, Hernandez-Contreras D, Peregrina-Barreto H, de Jesus Rangel-Magdaleno J, Ramirez-Cortes JM Deep learning classification for diabetic foot thermograms. *Sensors* 2020, 20(6): 1762. https://doi.org/10.3390/s20061762

Gavin JR, Alberti KGMM, Davidson MB, Defronzo RA, Drash A, Gabbe SG, Genuth S, Harris MI, Kahn R, Keen H, Knowler WC, Lebovitz H, Maclaren NK, Palmer JB, Raskin P, Rizza RA, Stern MP. Expert committee on the diagnosis and classification of diabetes mellitus: Report of the expert committee on the diagnosis and classification of diabetes mellitus. *Diabetes Care* 2003, 26(Suppl. 1): S5–S20.

Facchinetti A, Del Favero S, Sparacino G, Castle JR, Kenneth Ward W, Cobelli C Modeling the glucose sensor error. *IEEE Transactions on Biomedical Engineering* 2014, 61(3): 620–29. https://doi.org/10.1109/TBME.2013.2284023

Francis Ring EJ History of thermology and thermography: Pioneers and progress. *Thermology International* 2012, 22(3).

Harding JR, Wertheim DF, Williams RJ, Melhuish JM, Banerjee D, Harding KG Infrared imaging in diabetic foot ulceration. 2002, November, 916–918. https://doi.org/10.1109/IEMBS.1998.745591

Hernandez-Contreras D, Peregrina-Barreto H, Rangel-Magdaleno J, Gonzalez-Bernal JA, Altamirano-Robles L A quantitative index for classification of plantar thermal changes in the diabetic foot. *Infrared Physics & Technology* 2017, 81(March): 242–49. https://doi.org/10.1016/J.INFRARED.2017.01.010

Hernandez-Contreras DA, Peregrina-Barreto H, Rangel-Magdaleno De Jesus J, Renero-Carrillo FJ Plantar thermogram database for the study of diabetic foot complications. *IEEE Access* 2019, 7: 161296–161307. https://doi.org/10.1109/ACCESS.2019.2951356

Ilo A, Romsi P, Mäkelä J Infrared thermography and vascular disorders in diabetic feet. *Journal of Diabetes Science and Technology* 2020, 14(1): 28–36. https://doi.org/10.1177/1932296819871270

Amalu W International academy of clinical thermology quality assurance guidelines standards and protocols in clinical thermographic imaging. *Eagle Institute of Clinical Thermology* 2020, 2020.

Kaabouch N, Chen Y, Anderson J, Ames F, Paulson R Asymmetry analysis based on genetic algorithms for the prediction of foot ulcers. *Visualization and Data Analysis 2009* 2009, 7243 (January): 724304. https://doi.org/10.1117/12.805975

Karami E, Prasad S, Shehata M Image matching using SIFT, SURF, BRIEF and ORB: Performance comparison for distorted images. 2017, October. http://arxiv.org/abs/1710.02726

Lavery LA, Armstrong DG Temperature monitoring to assess, predict, and prevent diabetic foot complications. *Current Diabetes Reports* 2007, 7(6): 416–419. https://doi.org/10.1007/s11892-007-0069-4

Lavery LA, Higgins KR, Lanctot DR, Constantinides GP, Zamorano RG, Armstrong DG, Athanasiou KA, Mauli Agrawal C Home monitoring of foot skin temperatures to prevent ulceration. *Diabetes Care* 2004, 27(11): 2642–47. https://doi.org/10.2337/DIACARE.27.11.2642

Liu C, van Netten JJ, van Baal JG, Bus SA, van der Heijden F Automatic detection of diabetic foot complications with infrared thermography by asymmetric analysis. *Journal of Biomedical Optics* 2015, 20(2): 026003. https://doi.org/10.1117/1.JBO.20.2.026003

Lowe DG Distinctive image features from scale-invariant keypoints. *International Journal of Computer Vision 2004 60:2* 2004b, 60(2): 91–110. https://doi.org/10.1023/B:VISI.0000029664.99615.94

Raghu M, Zhang C, Kleinberg J, Bengio S. Transfusion: Understanding transfer learning with applications to medical imaging. 2019. https://research.google/pubs/pub48545/

Renero CFJ The thermoregulation of healthy individuals, overweight–obese, and diabetic from the plantar skin thermogram: a clue to predict the diabetic foot. *Diabetic Foot and Ankle* 2017, 8(1). https://doi.org/10.1080/2000625X.2017.1361298

Ring EFJ, Ammer K Infrared thermal imaging in medicine. *Physiological Measurement* 2012, 33(3): R33. https://doi.org/10.1088/0967-3334/33/3/R33

Rother KI. Diabetes treatment — Bridging the divide. *The New England Journal of Medicine* 2007, 356(15): 1499-1501. https://doi.org/10.1056/NEJMP078030

Rublee E, Rabaud V, Konolige K, Bradski G ORB: An efficient alternative to SIFT or SURF. *Proceedings of the IEEE International Conference on Computer Vision* 2011, 2564–2571. https://doi.org/10.1109/ICCV.2011.6126544

Tareen SAK, Saleem Z A Comparative Analysis of SIFT, SURF, KAZE, AKAZE, ORB, and BRISK. *2018 International Conference on Computing, Mathematics and Engineering Technologies: Invent, Innovate and Integrate for Socioeconomic Development, ICoMET 2018 - Proceedings 2018-January (April)*, 2018: 1–10. https://doi.org/10.1109/ICOMET. 2018.8346440

Shamim Hossain M, Muhammad G, Guizani N Explainable AI and mass surveillance system-based healthcare framework to combat COVID-I9 like Pandemics. *IEEE Network* 2020, 34(4): 126–32. https://doi.org/10.1109/MNET.011.2000458

Simonyan K, Zisserman A. VGG-16." ArXiv Preprint. 2014.

Thirunavukkarasu U, Umapathy S, Janardhanan K, Thirunavukkarasu R. A computer aided diagnostic method for the evaluation of Type II diabetes mellitus in facial thermograms. *Physical and Engineering Sciences in Medicine 2020 43:3* 2020a, 43(3): 871–88. https://doi.org/10.1007/S13246-020-00886-Z

Thirunavukkarasu U, Umapathy S, Krishnan PT, Janardanan K Human tongue thermography could be a prognostic tool for prescreening the Type II diabetes mellitus. *Evidence-Based Complementary and Alternative Medicine* 2020b, 2020(January): 1–16. https://doi.org/10.1155/2020/3186208

van Netten JJ, van Baal JG, Liu C, van der Heijden F, Bus SA Infrared thermal imaging for automated detection of diabetic foot complications. *Journal of Diabetes Science and Technology* 2013, 7(5): 1122–29. https://doi.org/10.1177/193229681300700504

Vardasca R, Magalhaes C, Mendes J Biomedical applications of infrared thermal imaging: current state of machine learning classification. *Proceedings* 2019, 27(1): 46. https://doi.org/10.3390/PROCEEDINGS2019027046

Vardasca R, Vaz L, Mendes J Classification and decision making of medical infrared thermal images. *Lecture Notes in Computational Vision and Biomechanics* 2018, 26: 79–104. https://doi.org/10.1007/978-3-319-65981-7_4

Vollmer M, Möllmann K-P *Infrared thermal imaging: fundamentals, research and applications*. 2nd edition, Wiley, 794 pages, 2018.

Zimmet PZ, Magliano DJ, Herman WH, Shaw JE Diabetes: A 21st century challenge. *The Lancet Diabetes & Endocrinology* 2014, 2(1): 56–64. https://doi.org/10.1016/S2213-8587(13)70112-8

7 Thermal Imaging in Detection of Fever for Infectious Diseases

Fever is one of the common symptoms of illness for communicable and infectious diseases such as severe acute respiratory syndrome (SARS) and coronavirus disease (COVID-19). With the outbreak of SARS in the year 2003 and the COVID-19 pandemic in 2020, the need for a fever screening system is irrefutable for effective handling of these health emergencies, especially in populous countries (J. Zhang, Liu, and Zhu 2021). Conventional body temperature-measuring devices such as clinical thermometers which can provide accurate measurement are often contact-based and cannot be used for fever screening of large groups of people (Piccinini, Martinelli, and Carbonaro 2021). Thermal imaging has now become a valuable tool in the rapid fever screening after the recent outbreak of COVID-19. This chapter explains the application of thermal imaging in the detection of fever in individuals in public places as well as illustrates the measurement protocol to be followed for the correct estimation of body temperature for proper containment of infectious diseases.

7.1 HUMAN BODY TEMPERATURE

Measuring body temperature for the detection of a suspected infection is widely practiced in medicine. The procedure has a long history and one of the first and well-documented studies was done by Wunderlich (P. A. Mackowiak and Worden 1994). In his seminal book, Wunderlich reported temperature measurements recorded in a large sample of subjects using thermometry available at that time and concluded that the normal temperature value of healthy individuals falls in the range of 36.2°C (97.2°F) to 37.5°C (99.5°F) based on the different diurnal cycle. It was also reported a mean temperature of 37.0°C (98.6°F) for healthy individuals and any temperature measurement beyond 38°C (100.4°F) was meant febrile. However, Wunderlich understood that the sample size of healthy subjects was rather small and the measurement procedure in healthy subjects sometimes lacked sufficient precision (Wunderlich 1871).

Prior to the discussion of the normal value of deep body temperature, some basic definitions of the thermoregulation system are provided to foster the understanding of fever. Thermoregulation aims to maintain body core temperature (T_b) within a narrow range. T_b is often considered to be the temperature of inner organs, particularly of the blood in the heart and the brain. However, the core is more of a

DOI: 10.1201/9781003245780-7

TABLE 7.1

Definition of Thermoregulatory States

	IUPS Thermal Commission (2001)	Romanovsky et al. (2005)
Normothermia	The condition of a temperature regulator when its core temperature is within ± 1 SD of the range associated with the normal postabsorptive resting condition of the species in the thermoneutral zone.	A state characterized by a "normal" T_b
Hyperthermia	The condition of a temperature regulator when its core temperature is above its range specified for the normal active state of the species.	A state or response characterized by a "higher-than-normal" T_b
Hypothermia	The condition of a temperature regulator when its core temperature is below its range specified for the normal active state of the species.	A state or response characterized by a "lower-than-normal" T_b

Source: Romanovsky et al., 2005, IUPS Thermal Commission, 2001

concept than a practical body site. Measurement of true core temperature can only be obtained by invasive methods, but for convenience, oral, axillary, and rectal sites have become standard for clinical contact thermometry.

The level of body core temperature is described by terms such as "normothermia," "hyperthermia," and "hypothermia." Definitions of these terms are shown in Table 7.1. Irrespective of labeling the regulated core temperature as "setpoint" or as "equilibrium (or balance) point, T_{eq}" (Romanovsky et al., 2005), definite ranges of temperature for normo-, hypo-, and hyperthermia are not available, although the 3rd edition of the *Glossary of Terms for Thermal Physiology* (IUPS Thermal Commission, 2001) defines the upper and lower limit of normal T_b by ± 1 standard deviation of the mean. The "thermoneutral zone" is defined by the range of ambient temperature at which temperature regulation is achieved only by control of sensible heat loss, i.e., without regulatory changes in metabolic heat production or evaporative heat loss.

Fever is a state in which T_{eq} is above the normal level of T_b or a response in which T_b is temporarily regulated above its normal level. Or in other words, fever develops when the setpoint for core body is shifting upward. The neural structures controlling the setpoint are located in the hypothalamus. As a consequence of a well-functioning thermoregulatory response, humans constrict their cutaneous blood vessels and start shivering in the first phase of fever. In the case of fever solving, vasodilation of skin vessels and sweat production occur, indicating the setpoint return to the pre-febrile level. The shift of the setpoint is caused by pyrogens, molecular compounds of exogenic or endogenic origin. This results from central thermoregulatory vascular endothelial cells and neurons via pyrogenic cytokines in local production of prostaglandin E2 (PGE2) (Roth & Blatteis 2014). Pyrogens per se do not induce heat production but change the threshold for the activation of thermo-effector responses.

Wunderlich's observations were widely accepted throughout the medical fraternity for the assessment of fever in individuals (Mackowiak, Wasserman, and Levine 1992). More than a century later, a group of scientists headed by Mackowiak, University of Maryland, School of Medicine, conducted a study with a convenient-sized group of volunteers and obtained the normal temperature value of healthy individuals. Their study involved 148 individuals of which 122 were men and 26 were women with ages in the range of 18–40 years. More than 700 recordings of oral temperatures were recorded during different times of the day. Table 7.2 shows some of the temperature data available publicly for illustration of the recorded measurements. Figure 7.1 is a representation of the temperature distribution of normal subjects provided in Table 7.2.

TABLE 7.2
Human Body Temperature (°F) of Healthy Individuals (n = 106)

98.6	98.6	98.0	98.0	99.0	98.4	98.4	98.4	98.4	98.6	98.6
98.8	98.6	97.0	97.0	98.8	97.6	97.7	98.8	98.0	98.0	98.3
98.5	97.3	98.7	97.4	98.9	98.6	99.5	97.5	97.3	97.6	98.2
99.6	98.7	99.4	98.2	98.0	98.6	98.6	97.2	98.4	98.6	98.2
98.0	97.8	98.0	98.4	98.6	98.6	97.8	99.0	96.5	97.6	98.0
96.9	97.6	97.1	97.9	98.4	97.3	98.0	97.5	97.6	98.2	98.5
98.8	98.7	97.8	98.0	97.1	97.4	99.4	98.4	98.6	98.4	98.5
98.6	98.3	98.7	98.8	99.1	98.6	97.9	98.8	98.0	98.7	98.5
98.9	98.4	98.6	97.1	97.9	98.8	98.7	97.6	98.2	99.2	97.8
98.0	98.4	97.8	98.4	97.4	98.0	97.0				

Source: Philip A. Mackowiak, Wasserman, and Levine 1992

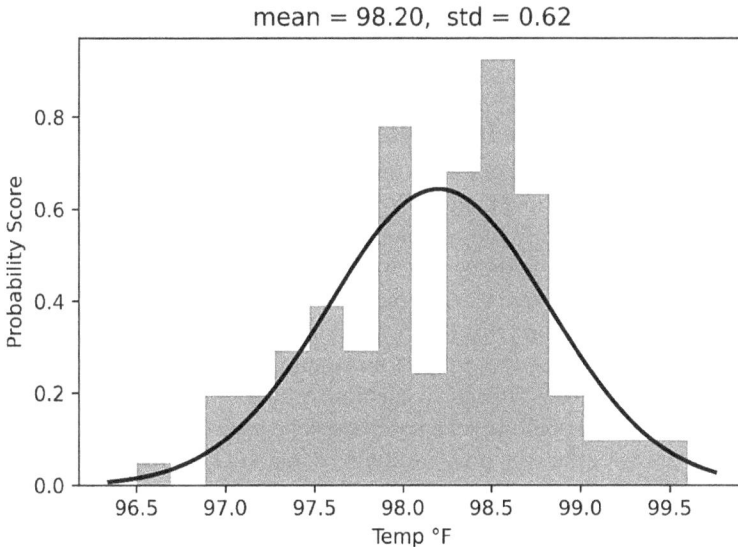

mean = 98.20, std = 0.62

FIGURE 7.1   Distribution of body temperature in a subsample of healthy individuals (n = 106) obtained from the work of Mackowiak et al. The 95% confidence interval of mean falls in the range of (98.08, 98.32).

It can be inferred from the work of Mackowiak et al. that 37.7°C (99.9°F) is regarded as the upper limit of oral temperature of healthy young adults and any value beyond this temperature may be regarded as suspicious of abnormality.

Waalen and Buxbaum (2011) found in a large cohort of 18630 white adults aged 20–98 years, a mean oral temperature of 97.3°F (36.3°C), with higher values in women 97.5 ± 1.2°F (36.4 ± 0.7°C) than in men 97.2 ± 1.1°F (36.2 ± 0.6°C). Mean oral temperature decreased with age in both men and women. Among women, temperatures were significantly lower in all age groups more than 40 years old compared with the 20–29-year-old age group. In men, differences were significant only when men older than 70 years were compared with 20–29 years old.

7.1.1 FEVER DEVELOPMENT DUE TO INFECTIOUS DISEASES

Elevated body core temperature (EBT > 37.7°C) is an important clinical sign suspicious of infection. Contact with pathogens such as viruses, bacteria, or parasites triggers the defense mechanisms of the immune system, resulting in the activation of tissue macrophages, leukocytes, and lymphocytes involved in the engulfing or digestion of the pathogens. Surface components of pathogens such as lipopolysaccharides (LPS) are currently the most studied and best understood exogenic pyrogens. LPS and other pathogen-associated molecular patterns (PAMP) bind to toll-like receptors of macrophages and dendritic cells resulting in the release of pyrogenic cytokines, e.g., interleukins IL-1, IL-6, tumor necrosis factor (TNF), and interferons (Wright & Auwaerter 2020). Thus often, EBT in humans suggests a possible viral or bacterial infection. Hence, detection of individuals with EBT is a necessary step in the management of infectious diseases such as SARS, MERS, and COVID-19 which are known to be contagious (El-Radhi 2018).

7.1.1.1 Thermal Imagers for Elevated Body Temperature Detection

Thermal scanners or detectors allow non-contact temperature measurement of the human body by infrared imaging by a process commonly referred to as thermal imaging. Unlike, infrared thermometers, thermal imagers provide a temperature distribution or thermal maps of the skin region under focus. Each pixel point in the thermal image is a radiometric value specifying the temperature of the measured skin region (Ng, Kawb, and Chang 2004). Figure 7.2 shows the facial thermograms – the temperature distribution of the test subject at various pixel points in the image. Image acquisition system of the thermal camera provides various pseudo coloring schemes that can be used to visualize the temperature maps. Some of the commonly available color palettes are (a) Gray level (b) White Hot (c) Black Hot (d) Iron Bow (e) Rainbow and also it varies with the camera manufacturer. Most thermal cameras have in-built measurement tools which can identify the peak temperature point in a frame of the thermal image. These could also be used to estimate the temperature of the person. However, care must be taken in proper selection of the measurement point for good temperature

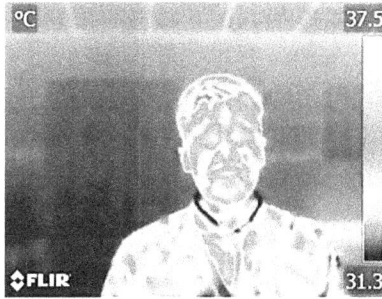

FIGURE 7.2 Facial thermogram of a volunteer. Here, the elevated temperature regions have color range toward the white color which could also be seen in the colorbar provided in the image. (See Umapathy and Krishnan 2021.)

measurement which represents an estimation of the actual core body temperature of the subject.

The selection of the type of thermal camera, installation of the device, the room setup, and measurement protocol are to be followed for correct temperature measurement and thus leading to proper fever detection. Some of the key specifications that ought to be considered for the selection of thermal camera for fever detection are as follows:

A. **Thermal Sensitivity**: It defines the ability of the thermal camera to detect the smallest change in the temperature variation when measuring a test subject. It is given in milliKelvin (mK) at a standard temperature condition. A value <50 mK is usually appropriate for temperature measurement in the detection of fever conditions. Thermal sensitivity has a large influence on the image quality of thermal images.

B. **Thermal Accuracy**: It specifies the deviation from the true temperature of the measured object. The typical uncertainty of infrared cameras is ± 2°C or ± 2% of the total temperature range. However, camera manufacturers claim in their specification lists accuracies between 0.3°C and 0.5°C, often for a temperature range between 15°C and 32°C. While this temperature band is different from the relevant temperatures of 18°C (image background) to 38°C (febrile subjects) occurring in fever screening, the accuracy data are not confirmed in an independent external evaluation.

Accuracy has to be determined by comparing camera readings against the temperature of a black body source. A quality assurance project conducted in England demonstrated that four of six investigated FLIR-cameras accurately measured the target temperature of 30°C but underestimated with stable increments in average temperature from 20°C to 30°C and overestimated temperatures from 30°C to 40°C. The limits of agreement between camera readings and black body temperature were found between −0.02°C to 0.69°C and −3.95°C to −2.41°C, the mean bias of all cameras was well within the manufacturer´s specification of ± 2 (5)°C (Marjanovic et al. 2018). However, the high uncertainty of the worst-performing camera questions its appropriateness for clinical applications.

C. **Range of Operation**: Thermal cameras have a lens system which is used to focus a particular region of interest of the object under consideration. Based on the focal length of the lens system, there is a limitation on the extent to which we can measure the temperature of the object from a distance. This is the range of operation of the thermal camera. It is provided in meters. Commonly used thermal cameras for fever detection have an operating range between a few decimeters and 30 m.

D. **Field of View (FOV)**: It specifies the amount of information the thermal camera can account for in a given frame of reference. This value is based on camera design parameters such as the focal length of the camera lens and the size of the sensor. The FOV value is provided as horizontal FOV (deg) × vertical FOV (deg).

E. **IR Resolution**: In a thermal camera, the sensor consists of an array of temperature-sensitive elements – the ones often used are uncooled bolometers with a size of a few micrometers. The number of such elements in the sensor array is defined as the resolution of the thermal camera. A standard thermal camera for fever detection needs a sensor array size of 320 × 240 which amounts to 76800 temperature measuring elements for a frame of the image.

Table 7.3 summarizes some of the available thermal imagers which could be used for temperature measurement of humans in fever detection.

7.1.2 MEASUREMENT PROTOCOL

Application of thermal imaging in fever detection or in mass fever screening requires a standard protocol for getting a reliable performance of the system, thus reducing the false positivity or negativity rate. Earlier works on the formation of a standard protocol for fever screening based on thermal imaging could be dated back to the onset of the SARS pandemic in China in the year 2002–2003 (Chiu et al. 2005). Many research works were carried out to determine the optimal measurement setup for proper fever screening using thermograms. After due deliberations and overall consensus in the international community, a standardized International Standards Organization (ISO) protocol was defined in "Particular requirements for the basic safety and essential performance of screening thermographs for human febrile temperature screening" part of IEC80601-2-59 and ISO subcommittee SC3. This document was published in September 2008 and provides guidelines for fever screening (Ring et al. 2013). Figure 7.3 shows a typical setup for fever screening of a person in a controlled environment. Figure 7.4 shows a modified illustration of a room setup for mass fever screening of people in more crowded places. The IR camera and the test subject from whom the measurement is made have to be in the same temperature-controlled room and the distance between the region of interest and the camera should be well within the

TABLE 7.3
Comparison of Thermal Imagers Available for Human Body Temperature Measurement

Manufacturer/Model	Temperature Range	Thermal Accuracy	Thermal Sensitivity	IR Resolution	FOV
FLIR Exx-Series	−20–120°C	± 2% of reading for ambient temperature 15–35°C	<40 mK @ 30°C	320 × 240 (FLIR E75) 464 × 348 (FLIR E95) 384 × 288 (FLIR E85)	24° × 18° (17 mm lens), 42° × 32° (10 mm lens), 14° × 10° (29 mm lens)
Optotherm Thermoscreen	0–80°C	± 0.3°C between 30°C and 40°C ± 1°C overall	<50 mK @ 25°C	640 × 480	48° × 38° (12 mm lens)
GUIDE IR236 IR Fever Warning System	−10–50°C	≤0.3°C (ambient temperature 16–32°C)	<40 mK	400 × 300	38° × 28° (9.7 mm lens)
T120H Fever Screening Thermal Camera	20–50°C	≤0.5°C (ambient temperature 25°C, target distance 1 m)	60 mK	120 × 90	50° × 38° (2.28 mm lens)
FLIR A320 Tempscreen	−20–120°C	±2% of reading	<40 mK @ 30°C	320 × 240	25° × 18.8° (18 mm lens)
InfiRay ITS II 300	0–60°C	±0.5°C @ target temp of 33–42°C	<40 mK @ 25°C	384 × 288	47° × 35.6° (7.8 mm lens)

Doran 2020, Khaksari et al. 2021

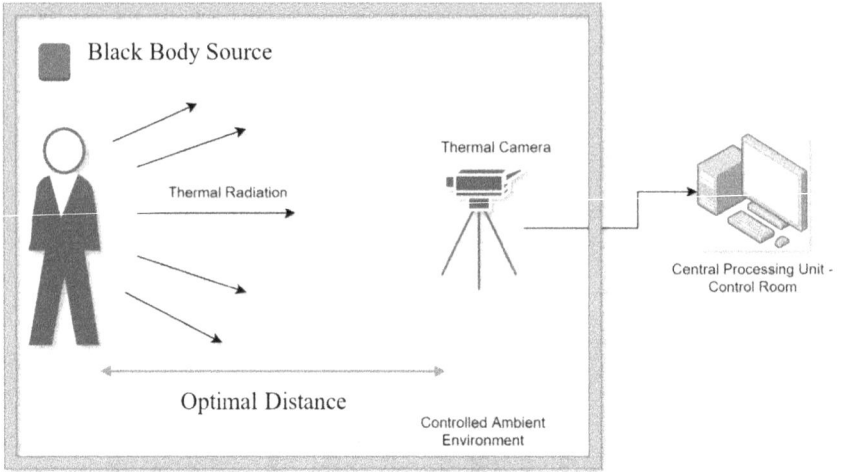

FIGURE 7.3 Fever screening setup using IR thermography. The measurement is taken in a controlled environment such that the temperature of the air is maintained at nominal conditions.

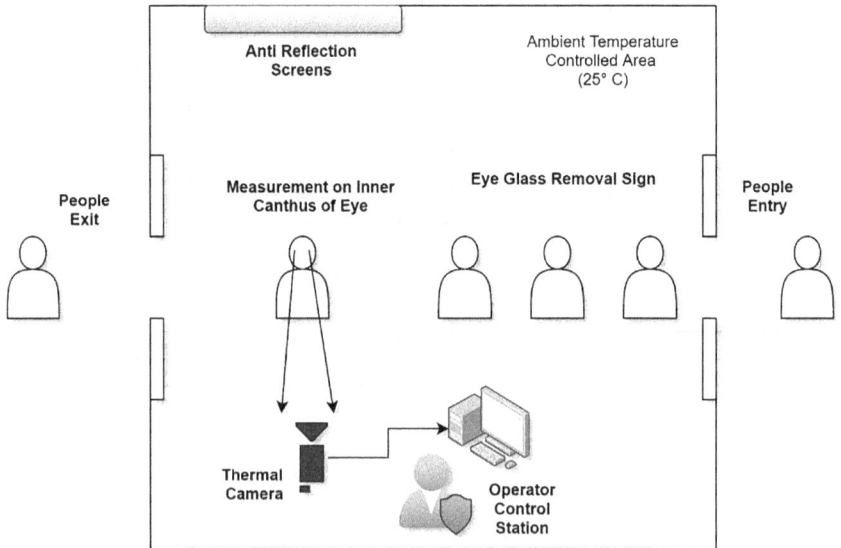

FIGURE 7.4 Mass fever screening setup using IR thermography. This schematic explains the setup for screening of multiple persons in a crowded place such as airports, malls, train stations, and other crowded areas.

range of the type of camera used for fever screening. Furthermore, the guidelines to be followed for correct fever screening are explained in the following section (Ghassemi et al. 2018):

A. **Room Temperature**: Measurement of human body temperature based on facial thermograms has to be done in a controlled environment where the effect of ambient temperature on the human body is minimal. The variation of temperature in the measurement station has to be restricted to around ± 1°C to minimize heat exchange between the human body and the ambiance.

B. **Patient Settling Time**: The constant heat exchange between objects and their environment leads finally to an unstable state of equilibrium. Any change in the heat content of this system will affect the heat transfer between elements within the system and thus their surface temperature. If surfaces are not in contact with other objects or oppose forced convection, heat dissipation occurs exclusively by infrared radiation, thus affecting infrared-based temperature measurement. Individuals moving from one thermal environment to another will alter their amount of heat transfer. The typical challenge is undressing, but also changing the ambient temperature by ~2°C will affect the uncovered facial skin temperature (see Chapter 2). Thus, an acclimation time is also required in case of fever screening with a thermal infrared camera. A minimum settling time of at least 15 minutes is recommended for an ambient temperature of around 24°C.

C. **Subject Preparation**: Temperature measurement based on facial thermal imaging requires the subject to be in a resting state or without any kind of exercise regime. The subject face is focused onto the FOV of the thermal camera such that the region of interest, namely the eye region is focussed properly on the camera. Care should be taken such that at least 3 × 3 pixels are present in the measurement region/eye region of the face. This requirement provides a reliable estimate of the body temperature of the subject. Care must be taken that eyeglasses are to be removed because glass act as both a reflective medium to thermal radiations and have low transmittance for infrared radiation with wavelengths between 6 and 15 μm. So, subjects who wear eyeglasses have to be instructed to remove eyeglasses before going for temperature measurement based on facial thermograms as shown in Figure 7.4.

D. **Measurement Site**: Selection of the region of interest for temperature measurement of core body temperature is a huge topic of debate. Many regions in the human body were studied in the literature for determining the correct region for acquiring the core body temperature of the subject. Often, temperature measurement in the forehead or in the skin region provides a different temperature reading when compared to deep body temperature (Ammer and Ring 2019). In clinical medicine, temperatures measured in the rectum or esophagus are considered to be the reliable estimate of true body temperature. However, these sites are not easily

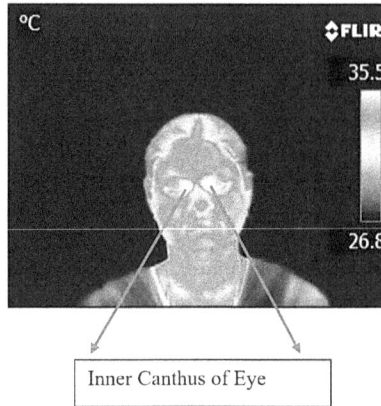

FIGURE 7.5 Preferred region of interest of facial thermograms for measuring the core body temperature for screening of fever condition.

accessible, the measurement is invasive and hence axillary and sub-lingual, are preferred for contact-based clinical temperature measurement. The temperature measured invasively provides a higher body temperature than obtained with contact thermometers. In the case of non-contact temperature measurements such as thermal imaging, much research is still needed to determine the relationship between the core body temperature and the selected skin region on the face. However, recently, a few researchers have studied this relationship and reported that the regions near the eye corners or the inner canthus of the eyes show a good relationship to body core temperature. This is also illustrated in Figure 7.5 which is a facial thermogram of a test subject. From the figure, we could observe that the temperature near the eye corners has a higher temperature compared to other facial regions.

It is important to remind the reader that screening and diagnosing are two different procedures. Nicholas J. Wald gave the following definition (1994): "Screening is the systematic application of a test or enquiry to identify individuals at sufficient risk of a specific disorder to warrant further investigation or direct preventive action, amongst persons who have not sought medical attention on account of symptoms of that disorder." Subjects who proved positive in a screening test must be confirmed as positive cases by a standard diagnostic test. Threshold values for positive cases depend on the applied method of measurement. While the threshold for fever is 38°C for rectal temperature (Niehues, 2013), 37.7°C was reported for oral temperature (Mackowiak, Wasserman, and Levine 1992), and 37.8°C for ear temperature (Cho and Yoon, 2014). The difference of oral and ear temperature to rectal temperature builds the rationale for different fever thresholds. The selection of the threshold for screening should be determined by the rate of false-negative cases.

7.2 FEVER DETECTION SYSTEMS

Fever is one of the common clinical signs of an underlying condition of a person which may be caused by an infectious disease but inflammatory, neoplastic diseases, return from traveling, and miscellaneous others must be considered as fever causes (Wright & Auwaerter 2020). Though fever could not be directly related to the causal physiological problem, often secondary tests such as blood, saliva, or imaging analysis are done to correctly identify the underlying problem in the febrile person. Nevertheless, screening individuals for signs of fever is a necessary step in restricting the spread of infectious diseases such as SARS or COVID-19. Applications of fever screening systems for containing infectious disease outbreak could be traced back to the year 2003, where it was originally conceptualized by Singapore's Defence Science and Technology Agency (DSTA) and Singapore Technologies Electronics for screening of individuals in airports during the SARS outbreak (Chiu et al. 2005). The method involved using thermal cameras to identify individuals with EBT based on a few measurement points on the face followed by separate clinical measurements of detected febrile cases. Since then, it has become a de facto standard for screening a large number of people in crowded places such as airports and community centers.

In this section, we review some of the conventional methods of screening individuals for EBT and some of the recent trends of using artificial intelligence in detecting fever conditions in a person.

7.2.1 CONVENTIONAL METHODS

Measurement of elevated human body temperature is done by using various commercially available clinical equipment which could be classified as contact and non-contact temperature measuring devices. Conventional contact devices are based on temperature-sensitive metallic strips or a heat-sensing fluid such as mercury or alcohol. These are also called clinical thermometers which are usually placed in the armpit or under the tongue for correct temperature measurement. In some cases, these devices could also be used to measure temperature in the rectal region of the subject. Often, the rectal temperature reading is around 0.6°C higher than the oral-based readings and is also considered more reliable than the latter. Other classes of conventional thermometers are based on non-contact temperature measurement that uses infrared technology to determine the human body temperature. Two of the widely used types are the forehead and tympanic infrared thermometers. These devices rely on measuring the body temperature based on converting the radiated thermal energy from the human skin into electrical energy which is then digitized to provide a digital readout of the temperature. Two types of forehead IR thermometers are available in the market (Martin et al., 2021):

1. The so-called temporal artery thermometer is intended to capture the infrared radiation emitted from the region of the superficial temporal artery. It is placed touching the temple and it has a cup shape at the end. Generally, it is

moved over the skin surface from the center of the forehead toward the ear and the maximum temperature is recorded.

2. The more general forehead thermometer measures at just one point, several centimeters from the skin.

Tympanic thermometers measure the radiant energy in the ear canal. In children, the mean difference between tympanic scan, forehead skin scan, and temporal artery scan compared with rectal measurement was 0.49°C, 0.34°C, and 0°C, respectively. All tools overestimated the temperature at lower body temperatures and underestimated at higher body temperatures (Allegaert, et al. 2014).

Zhou et al. 2020 employed an infrared camera for comparing the temperature of various face regions to oral temperature serving as the reference. For each facial region in the thermal image, comparative boxplots of pairwise differences with reference temperature were constructed. The findings indicated that forehead temperatures provide a generally inferior estimation of reference (oral) temperature relative to the inner canthi and full-face maximum temperature.

The Validity and reliability of the temperature measurement based on clinical thermometers for fever detection rely on the selection of the measurement site (Piccinini, Martinelli, and Carbonaro 2021; Ammer and Ring 2019). Studies have shown that often rectal and tympanic temperature measurements are a good approximation of the true body core temperature. However, the application of these procedures is not practical and is time-consuming when designing fever screening systems in crowded places (Ring et al. 2013; Ammer and Ring 2019). Hence, more advanced temperature measuring systems based on thermal imaging are widely used for mass fever screening in public places (Zhang, Liu, and Zhu 2021; Piccinini, Martinelli, and Carbonaro 2021). In the following sections, we discuss some of the artificial intelligence-based fever screening systems based on facial thermography.

7.2.2 AI-BASED FEVER SCREENING SYSTEMS

Facial thermography for fever detection is a non-contact-based temperature measurement system that relies on determining the core body temperature based on the pre-select measurement site on the facial region of the subject. The act of determining the human body temperature based on facial thermograms is illustrated in Figure 7.6, where a thermal camera is deployed for acquiring a continuous stream of facial images. Recent studies show that temperature in the inner canthi of the human eyes is a good approximation of the true body core temperature. With the availability of high computing power and dedicated hardware for training complex AI algorithms, a fully automated mass fever screening system based on facial thermograms is widely researched (Ring et al. 2013; Perpetuini et al. 2021).

The heart of the AI-based fever screening system involves dedicated machine learning and deep learning algorithms for face detection, after which the facial markers, especially the eye corners, are extracted and processed for temperature

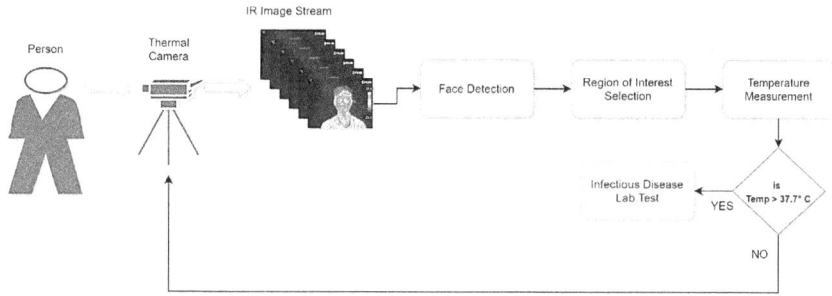

FIGURE 7.6 AI-based continuous temperature measurement of humans for detection of fever.

measurement. Once the temperature at the inner eye corners is obtained from the facial thermograms, the EBT is checked based on the pre-defined condition of a threshold value of around 37.7°C. A person having a temperature value greater than this pre-defined and widely accepted value suggests a febrile condition. Since fever is a common symptom for many underlying conditions in the human body, further tests are initiated for the detection of infectious diseases such as SARS, MERS, or COVID-19. A widely followed protocol for persons suggestive of having a febrile condition is following a quarantine procedure with contact tracing of the subject. The AI-based fever screening system could be an effective tool in containing the disease spread owing to its speed of operation.

7.2.2.1 Face and Eye Detection

The principal step in developing an automated fever screening system is the detection of facial landmarks and extracting the key facial features such as the eye corners. This is accomplished by first detecting faces in the input thermograms. Detection of human faces is a widely researched topic that is used in various fields for human face recognition. Over the decades, many machine learning and deep learning algorithms were developed for the detection of faces in the input images. In the following section, we explain the methods and procedures for finding faces in the images.

7.2.2.1.1 Machine Learning

Face detection from a video feed or from a stream of images is an increasingly used application in many fields such as crowd monitoring, security surveillance, crowd flow prediction, and face recognition. Very recently, face detection from thermal imagery is used in identifying febrile subjects based on EBT from the eye regions. In this section, we discuss one of the widely used methods of face detection which was originally proposed by Viola and Jones in 2001 (Viola and Jones, n.d.) and further researched by authors such as Wang (2014) and Huang, Shang, and Chen (2019). Due to its computational efficiency and speed of operation, it is one of the widely used methods for face detection from images and is also used in

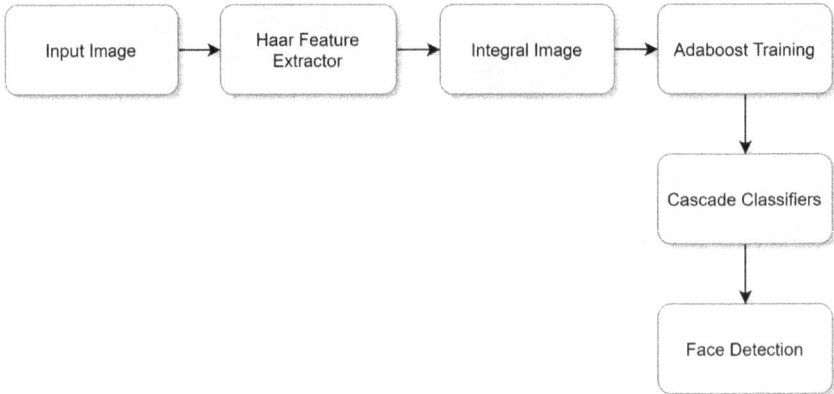

FIGURE 7.7   Face detection based on machine learning classifier and Haar-like feature extractor. (See Wang 2014.)

face detection in video signal. Figure 7.7 shows the key components that make up the algorithm for face detection from input images.

The procedure for finding the face in an image is based on dividing the whole image into smaller sub-images. Then, from each sub-image, Haar-like features are extracted and provided as input to a machine learning classifier such as the AdaBoost. The classifier based on the extracted features performs a binary classification of whether the sub-image corresponds to a human face or not.

A. **Haar-Like Features**: Extraction of facial markers such as the eyes, nose, and mouth using a set of textural features provide a computationally efficient procedure when compared to pixel-based face detection. Figure 7.8 shows some of the feature extractors for the detection of facial markers. These Haar-like features resemble the facial contours and play a vital role in classifying the face in the image. Considering a pixel dimension of 24×24 for the face detector mask, around 180000 Haar-like features could be computed for a sample input image of 384×288 pixels.

The computation of the features could be further explained as follows: For example, Figure 7.8A measures the difference in pixel intensities covering the brighter and the darker regions of the basis function. Similarly, Figure 7.8B computes the difference in the overlapping horizontal sections of the image and the feature mask. Figure 7.8C subtracts the intensity of the darker regions from the sum of the brighter regions. Finally, Figure 7.8D computes the pixel intensities from the sum of the difference between the diagonal sections of the feature mask. The feature values obtained by this approach are calculated for the input images and are used to train a machine learning classifier such as an AdaBoost classifier. However, the number of features to be computed exceeds the number of pixels of the sample input image. Hence, in the seminal paper by Viola and Jones, they proposed

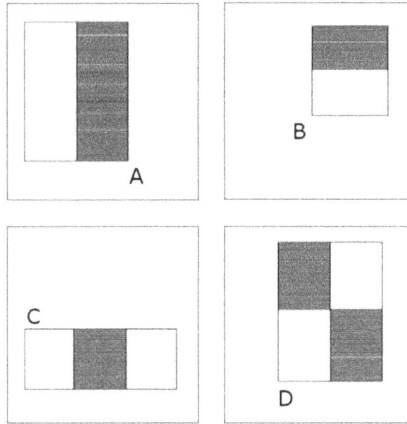

FIGURE 7.8   Haar-like features used in the work of Viola–Jones for detecting faces in the image.

a cumulative and fast feature computation method based on the integral image. This is explained in the following section.

B. **Integral Image**: Haar-like features as shown in Figure 7.8 are an over-complete set of image features. Computation of all these features takes a considerable computation cost. To speed up the feature extraction process, an integral image is formed which takes the dimension of the original image. However, the intensity values of pixels to the left and above the reference pixel are summed. This could also be represented by the following equation:

$$I\left(m,\ n\right) = \sum_{x \leq m, y \leq n} f\left(x,y\right) \tag{7.1}$$

Here, $f(x, y)$ represents the input image and $I(m, n)$ is the integral image with "m" and "n" being the reference pixel position. The integral image is an intermediate image that is used for feature extraction in such a way that the computation of rectangular features such as the Haar-like features shown in Figure 7.8 takes less time compared to actual pixel input images. This speed of operation is obtained as only four reference points are only required for the computation of rectangular features when computed from the integral image. After feature extraction, a machine learning classifier is trained on the features for face detection.

C. **AdaBoost Classifier**: A feature set is formed for the face (positive samples) and non-face (negative samples). The obtained feature set is high-dimensional data as the number of features is more compared to the number of pixels of the input images. For an acceptable performance of the face detector, only a few numbers of features are required. The selection of best features for detecting the face is an important task. This is achieved by using an

AdaBoost classifier. The training process starts with a weak classifier and is trained on only one feature at a time. Later, numerous such weak classifiers are trained for different feature values and the classifiers which yield prediction results above a predefined threshold value are selected in this stage. Later, the selected classifiers are combined for face detection. The procedure for training a series of classifier is illustrated in Equation 7.2. The final strong classifier is an aggregate of the "T" weak classifiers (Huang, Shang, and Chen 2019).

$$\hat{h}(x) = \begin{cases} 1 & \sum_{t=1}^{T} \log \frac{1}{\xi_t} h_t(x) \geq \frac{1}{2} \sum_{t=1}^{T} \frac{1}{\xi_t} \\ 0 & \text{otherwise} \end{cases} \tag{7.2}$$

Here, "ξ" is the error function of the weak classifier "$h(x)$" and the outcome of the strong classifier $\hat{h}(x)$ is face (Positive-"1") if the aggregate value exceeds the threshold limit for face given by $\frac{1}{2} \sum_{t=1}^{T} \log \frac{1}{\xi_t}$.

In the experiment done by Viola–Jones, they reported that only 6000 features are required for accurate face detection. This number is very less compared to the original feature size of around 180000. However, still, the number of selected features is more and inhibits the application of face detection in real-time video analysis where the frame rate is more than 15 fps. Thus, to overcome this problem, a cascade classifier is suggested.

D. **Cascade Classifier**: The AdaBoost classifier which was trained on the face and non-face images with acceptable prediction probability is selected based on a pre-defined threshold value. AdaBoost is an ensemble classifier which involves training a series of weak classifiers based on each feature. Then, the final prediction is based on the aggregate sum of the individual weak classifiers. Though, the ensemble approach of the AdaBoost classifier provides superior performance, applying it for real-time processing of videos is time-consuming. A possible solution to run the classifier in a concurrent way is using a cascade classifier method. This system places the different trained classifier systems in a cascade. It is an intuitive approach where the sub-images are provided as input to the first stage of the trained classifier. If the classifier output of the first stage is non-face, then the later stages of the cascade classifier are not run. The successive activation of the later stages of the cascade system depends on the previous stage. This approach provides fast processing of the input images for face or non-face as only fewer stages of the cascade classifier are run at any particular time. This immensely reduces the computation time of the face detection system as the probability of occurrence of a face in any image is comparatively less.

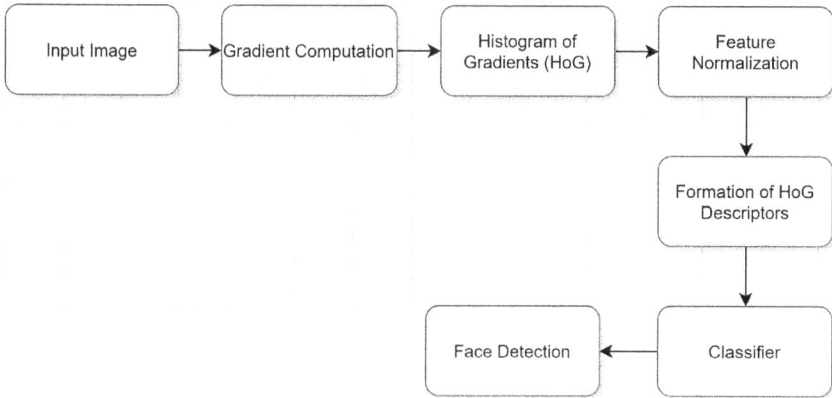

FIGURE 7.9 Face detection based on machine learning classifier and Histogram of Gradient feature extractor.

E. **Eye Extraction**: A combination of Haar-like features extracted from an integral image followed by a cascade classifier system provides a powerful face detection system for images. The system could also be used to detect facial markers such as eye, nose, and mouth based on the computed Haar-like features. In the fever detection system, the thermal images are given to face detection systems such as the Haar-cascades for first detecting the face, followed by detecting eye markers such as the corner of the eyes (inner canthus) for temperature measurement.

Similar to Haar-cascades which is explained in the above section, many other types of facial feature extractors were also widely investigated. Histogram of Gradients (HoG) is one such feature extractor that is used for face detection (Déniz et al. 2011). Figure 7.9 illustrates the various components involved in face/eye detection based on HoG feature extraction.

A. **Gradient Computation**: HoG is a class of pixel-level feature descriptors that relies on the computation of the gradient of each pixel in a pre-defined block of the image. Generally, a block size of 8×8 is used for HoG feature extraction. The method involves computing the gradient of the image in the horizontal and the vertical directions. This could be illustrated as follows (Shu, Ding, and Fang 2011):

$$g_x = \frac{\delta f(x,y)}{\delta x} \tag{7.3}$$

$$g_y = \frac{\delta f(x,y)}{\delta y} \tag{7.4}$$

$$G = \sqrt{g_x^2 + g_y^2} \qquad\qquad (7.5)$$

$$\theta = \tan^{-1}\left(\frac{g_x}{g_y}\right) \qquad\qquad (7.6)$$

Here, g_x represents the gradient vector in the x-direction, g_y represents the gradient in the y-direction, G is the magnitude of the gradient in (x, y) directions, and θ is the gradient orientation.

B. **Histogram of Gradients**: The obtained gradients are transformed into a histogram by means of dividing the gradient angle into nine bins (each bin has an angle range of 20°). Then, each gradient magnitude which falls within the corresponding bin is proportionately counted to form the HoGs. For example, if a gradient magnitude of 50 with an orientation angle of the pixel is 30°, then the magnitude is equally distributed such as 25 for bin 20 and 25 for bin 40 as 30° is midway between the bins at 20° and 40°, respectively. A key feature of this computation is that the HoG vector is translation invariant.

C. **Feature Normalization**: In order to make the HoG feature descriptors robust to variance in the lighting conditions of the image, a block-wise normalization is done. This can be achieved by considering the four rectangular neighborhood 8 × 8 image blocks leading to a block size of 16 × 16.

D. **Classifier**: After HoG normalization, a feature vector is formed by concatenating the normalized image gradients blocks into a single high dimensional feature vector. This procedure is done for a number of face and non-face images and the HoG feature descriptors are formed. A machine learning classifier such as the support vector machine is trained on the extracted HoG feature sets and could be used for the classification of the image block as either face or non-facial image.

7.2.2.1.2 Deep Learning Models

Face detection by deep learning models is widely researched owing to the easy availability of computing power which can provide accelerated computing of model weights during the training of the models. Convolution Neural Network (CNN) is a class of deep learning models that is widely used in computer vision. A recent survey of related work which involves face detection from images shows that there is a plethora of CNN architectures available for face detection (Jiang and Learned-Miller 2017). However, a multi-task cascade convolutional neural network (MTCNN) is one such CNN architecture that has been reported with state-of-the-art results. In this section, we cover the architecture design of this network and learn about the hyperparameters for proper tuning of the system for face detection from images (Yang et al. 2015; Chunming and Zhang 2021).

A. **CNN Architecture**: MTCNN consists of cascaded CNN architectures, namely (i) Proposal net (P-net), (ii) Refine net (R-net), and (iii) Output net (O-net). Each network is a standalone CNN which starts coarsely in identifying the face in the image and gradually, the complexity of the CNN architecture is increased for reliable detection of face in the input image (K. Zhang et al. 2016). The multilayer architecture is shown in Figure 7.10 and the detailed architecture specifications are provided in Table 7.4. The number of convolution layers and the complexity of the network increase gradually at each stage. The P-net has three convolution layers and the final layer of this stage provides the output for face classification (binary classification – face or non-face), the bounding box of the face regions, and the identification of the facial landmarks. This stage is a shallow CNN for the coarse identification of facial regions in the input region. It identifies a lot of probable regions which consist of facial features. However, the network produces many false positives from the given input image.

The R-net or the Refine CNN network is a deeper architecture with three convolution layers, two max-pooling layers, and a fully connected layer with 128 neurons. The main purpose of the R-net is to refine the detected facial regions by rejecting the probable non-face regions. Moreover, the bounding box of the detected facial regions and facial landmarks are formulated as a regression problem, and the coordinates of the bounding box and facial landmarks are updated iteratively. Finally, the O-net is the deepest of the CNN stages in the MTCNN that consists of four convolution layers, three max-pooling layers, and a fully connected layer of 256 neurons. This stage provides the finest classification accuracy owing to its deeper and more complex CNN architecture.

B. **Training of MTCNN**: Multi-task CNN involves multiple stages which are trained differently based on the task it is performing. There are three tasks which are done based on the loss function: (i) the classification of the input sub-image as either face or non-face, (ii) generation of the face bounding box, and (iii) generation of facial landmarks of the face images. The training of these tasks is carried by different loss functions which are provided in Equations (7.7–7.9). The first task is the classification of sub-images into "face" or "non-face," which is a binary classification problem that involves training the CNN stages based on the cross-entropy loss function as shown in Equation 7.7.

$$J_i = -\left(y_i \log\left(p_i\right) + \left(1 - y_i\right)\left(1 - \log\left(p_i\right)\right) \right) \qquad (7.7)$$

Here, J is the cross-entropy function with p_i being the prediction probability of the CNN model and $y_i \in \{0, 1\}$ being the ground truth of the training sample input (Du 2020).

Equation 7.8 specifies the mean-squared loss function for the bounding box of the CNN stages. Here, the estimation of the bounding box around the facial region in the input sub-image is formulated as a regression problem. L is the error

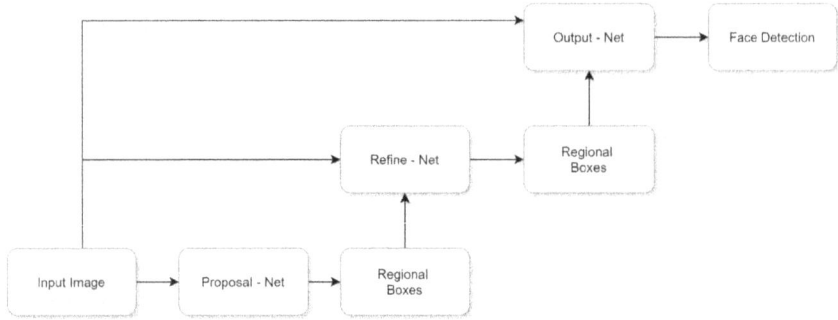

FIGURE 7.10   Face detection based on deep learning classifier and multi-cascade convolution neural network.

function with $\hat{y}^b$ being the model output and y^b is the ground truth of the facial coordinates of the input image.

$$L_i^b = \|\hat{y}_i^b - y_i^b\|_2^2 \tag{7.8}$$

Equation 7.9 describes the error function for the facial landmarks such as the eye, nose, and mouth coordinates of the model output $\hat{y}^p$ and ground truth y^b. Similar to the bounding box, the estimation of the facial landmark coordinates is stated as a regression problem (Ma and Wang 2018).

$$L_i^p = \|\hat{y}_i^p - y_i^p\|_2^2 \tag{7.9}$$

During each iteration, the model parameters are tuned in such a way that the loss functions are minimized for proper face detection from input images. Though the training of the CNN model requires a dedicated GPU for accelerated computing of the model weights, they provide state-of-the-art results in the detection of faces in images. The method could be applied for the detection of facial features from thermal images for fever screening.

7.3   CONCLUSION

Thermal imaging is a useful tool for mass screening of people in the detection of EBT or febrile conditions. Thermal cameras owing to their simple setup and speed of operation, provide a reliable screening for febrile conditions in people and thus limit the use of lab tests only to suspected cases. This extensively reduces the scanning time of a wide population, especially when the thermal imaging system is deployed in crowded places such as airports and public areas but only under strictly controlled measurement conditions. In this chapter, we have discussed the physiological significance behind the onset of fever in person due to infectious

TABLE 7.4
Multi-task Cascade CNN and Its Architecture Design for Face Detection

CNN Group	Input Layer Size	Convolution Layer Filter Size	Output Layer Size
P-Net	$12 \times 12 \times 3$	Stage 1: Conv: 3×3 MP: 2×2 Stage 2: Conv: 3×3 Stage 3: Conv: 3×3	Face classification: $1 \times 1 \times 2$ Bounding box: $1 \times$ 1×4 Facial landmark: 1 $\times 1 \times 10$
R-Net	$24 \times 24 \times 3$	Stage 1: Conv: 3×3 MP: 3×3 Stage 2: Conv: 3×3 MP: 3×3 Stage 3: Conv: 2×2 FCN: 128	Face classification: 2 Bounding box: 4 Facial landmark: 10
O-Net	$48 \times 48 \times 3$	Stage 1: Conv: 3×3 MP: 3×3 Stage 2: Conv: 3×3 MP: 3×3 Stage 3: Conv: 3×3 MP: 2×2 Stage 4: Conv: 2×2 FCN: 256	Face classification: 2 Bounding box: 4 Facial landmark: 10

"MP" refers to the max-pooling layer, "Conv" refers to the convolution layer, and "FCN" represents the fully connected layer. (See Zhang et al. 2016.)

diseases and the role of thermal imaging system in detecting the EBT for fever screening. We have also provided resources regarding some of the widely available thermal imagers for fever screening purposes. Moreover, we have covered in detail, the measurement protocol for correct screening of people for reliable performance in detecting fever. Besides, we have also discussed the recent trends in applying face/eye detection algorithms based on machine learning and deep learning in automated screening of fever conditions.

ACKNOWLEDGMENT

This work is supported by Science and Engineering Research Board (SERB), Department of Science and Technology, India, through the Core Research Grant (File No. CRG/2020/003042).

REFERENCES

Ammer K, Ring F *The Thermal Human Body. The Thermal Human Body.* Jenny Stanford Publishing, 2019. https://doi.org/10.1201/9780429019982

Chiu WT, Lin PW, Chiou HY, Lee WS, Lee CN, Yang YY, Lee HM, et al. Infrared thermography to mass-screen suspected SARS patients with fever. *Asia-Pacific Journal of Public Health* 2005, 17(1): 26–28. https://doi.org/10.1177/101053950501700107

Cho KS, Yoon J Fever screening and detection of febrile arrivals at an international airport in Korea: Association among self-reported fever, infrared thermal camera scanning, and tympanic temperature. *Epidemiology and Health* 2014, 36(1): e2014004.

Chunming Wu, Zhang Y MTCNN and FACENET based access control system for face detection and recognition. *Automatic Control and Computer Sciences* 2021, 55(1): 102–12. https://doi.org/10.3103/S0146411621010090

Déniz O, Bueno G, Salido J, De la Torre F Face recognition using histograms of oriented gradients. *Pattern Recognition Letters* 2011, 32(12): 1598–1603. https://doi.org/10.1016/j.patrec.2011.01.004

Doran G Introduction to temperature screening principles - Irish manufacturing research. *Imr.Ie* 2020. imr.ie/2020/05/20/Comparison-of-Temperature-Screening-Technologies-IMR-v3.pdf

Du J High-precision portrait classification based on MTCNN and its application on similarity judgement. *Journal of Physics: Conference Series* 2020, 1518(April): 012066. https://doi.org/10.1088/1742-6596/1518/1/012066

El-Radhi AS Fever in common infectious diseases. In: *Clinical Manual of Fever in Children*, Springer International Publishing, 2018, pp. 85–140. https://doi.org/10.1007/978-3-319-92336-9_5

Ghassemi P, Joshua Pfefer T, Casamento JP, Simpson R, Wang Q Best practices for standardized performance testing of infrared thermographs intended for fever screening. *PLoS One* 2018, 13(9): e0203302. https://doi.org/10.1371/journal.pone.0203302

Huang J, Shang Y, Chen H Improved viola-jones face detection algorithm based on HoloLens. *Eurasip Journal on Image and Video Processing* 2019, 2019(1): 1–11. https://doi.org/10.1186/s13640-019-0435-6

IUPS Thermal Commission Glossary of terms for thermal physiology. *The Japanese Journal of Physiology* 2001, 51(2): 245–280.

Jiang H, Learned-Miller, E Face detection with the faster R-CNN. *2017 12th IEEE International Conference on Automatic Face & Gesture Recognition (FG 2017)*, 2017, 650–657. IEEE. https://doi.org/10.1109/FG.2017.82

Khaksari K, Nguyen T, Hill B, Quang T, Perreault J, Gorti V, Malpani R et al Review of the Efficacy of Infrared Thermography for Screening Infectious Diseases with Applications to COVID-19. *Journal of Medical Imaging (Bellingham, Wash.)* 2021, 8(Suppl 1): 010901. https://doi.org/10.1117/1.JMI.8.S1.010901

Ma M, Wang J Multi-View Face Detection and Landmark Localization Based on MTCNN. *2018 Chinese Automation Congress (CAC)*, 2018, 4200–4205. IEEE. https://doi.org/10.1109/CAC.2018.8623535

Mackowiak PA, Wasserman SS, Levine MM A critical appraisal of 98.6°F, the upper limit of the normal body temperature, and other legacies of carl reinhold august Wunderlich. *The Journal of the American Medical Association* 1992, 268(12): 1578–80. https://doi.org/10.1001/jama.1992.03490120092034

Mackowiak PA, Worden G Carl Reinhold august wunderlich and the evolution of clinical thermometry. *Clinical Infectious Diseases* 1994, 18(3): 458–458. https://doi.org/10.1093/clinids/18.3.458

Marjanovic EJ. Britton J, Howell KJ, Murray AK. Quality assurance for a multi-centre thermography study. *Thermology International* 2018, 18(1): 7–13.

Martin MJ Knazovicka L, McEvoy H, Machin G, Pusnik I, Cardenas D, Sadli M, Chengdu B, Li W, Saunders P, Girard F Best practice guide: Use of infrared forehead thermometers to perform traceable non-contact measurements of human body temperature. Accessed on 2 December 2021. https://www.researchgate.net/publication/353947185_BEST_PRACTICE_GUIDE_USE_OF_INFRARED_FOREHEAD_THERMOMETERS_TO_PERFORM_TRACEABLE_NON-CONTACT_MEASUREMENTS_OF_HUMAN_BODY_TEMPERATURENg

Ng EYK, Kawb GJL, Chang WM Analysis of IR thermal imager for mass blind fever screening. *Microvascular Research* 2004, 68(2): 104–9. https://doi.org/10.1016/j.mvr.2004.05.003

Niehues T The febrile child: diagnosis and treatment. *Deutsches Arzteblatt International* 2013, 110(45): 764–74.

Perpetuini D, Filippini C, Cardone D, Merla A An overview of thermal infrared imaging-based screenings during pandemic emergencies. *International Journal of Environmental Research and Public Health* 2021, 18(6): 1–12. https://doi.org/10.3390/IJERPH18063286

Piccinini F, Martinelli G, Carbonaro A Reliability of body temperature measurements obtained with contactless infrared point thermometers commonly used during the COVID-19 pandemic. *Sensors* 2021, 21(11): 3794. https://doi.org/10.3390/s21113794

Ring EFJ, Jung A, Kalicki B, Zuber J, Rustecka A, Vardasca R New standards for fever screening with thermal imaging systems. *Journal of Mechanics in Medicine and Biology* 2013, 13(2). https://doi.org/10.1142/S0219519413500450

Romanovsky AA, Almeida MC, Aronoff DM, Ivanov AI, Konsman JP, Steiner AA, Turek VF Fever and hypothermia in systemic inflammation: Recent discoveries and revisions. *Front Bioscience* 2005, 10(1–3): 2193–2216.

Roth J, Blatteis CM Mechanisms of fever production and lysis: lessons from experimental LPS fever. *Comprehensive Physiology* 2014, 4: 1563–604.

Shu C, Ding X, Fang C Histogram of the oriented gradient for face recognition. *Tsinghua Science and Technology* 2011, 16(2): 216–24. https://doi.org/10.1016/S1007-0214(11)70032-3

Umapathy S, Krishnan PT Automated detection of orofacial pain from thermograms using machine learning and deep learning approaches. *Expert Systems*, 2021, 8(7), e12747. https://doi.org/10.1111/exsy.12747

Viola P, Jones, M Rapid Object Detection Using a Boosted Cascade of Simple Features. *Proceedings of the 2001 IEEE Computer Society Conference on Computer Vision and Pattern Recognition. CVPR 2001* n.d., 1:I-511–I-518. *IEEE Computer Socilogy.* https://doi.org/10.1109/CVPR.2001.990517

Waalen J, Buxbaum JN Is older colder or colder older? The association of age with body temperature in 18,630 individuals. *Journal of Gerontol A Biology Science Medicne Science* 2011, 66A(5): 487–492.

Wald NJ Guidance on terminology. *Journal of Medicine Screen* 1994, 1: 76.

Wang Y-Q An analysis of the viola-jones face detection algorithm. *Image Processing On Line* 2014, 4(June): 128–48. https://doi.org/10.5201/ipol.2014.104

Wright WF, Auwaerter PG Fever and fever of unknown origin: Review, recent advances, and lingering dogma. *Open Forum Infectious Diseases* 2020, 7(5): 132.

Wunderlich CA *On the Temperature in Diseases. A Manual of Medical Thermometry.* Translated from the Second German Edition by W Bathurst, MD Woodman, The New Sydenham Society, London, 1871.

Yang S, Luo P, Loy C-C, Tang X From Facial Parts Responses to Face Detection: A Deep Learning Approach. *2015 IEEE International Conference on Computer Vision (ICCV)*, IEEE, 2015, pp. 3676–84. https://doi.org/10.1109/ICCV.2015.419

Zhang J, Liu S, Zhu B Fever screening methods in public places during the COVID-19 pandemic. *Journal of Hospital Infection* 2021, 109(March): 123–24. https://doi.org/10.1016/j.jhin.2020.11.011

Zhang K, Zhang Z, Li Z, Qiao Y Joint face detection and alignment using multitask cascaded convolutional networks" *IEEE Signal Processing Letters* 2016, 23(10): 1499–1503. https://doi.org/10.1109/LSP.2016.2603342

Zhou Y, Ghassemi P, Chen M, McBride D, Casamento JP, Joshua Pfefer T, Wang Q Clinical evaluation of fever-screening thermography: Impact of consensus guidelines and facial measurement location. *Journal of Biomedical Optics* 2020, 25(09): 97002–97003. https://doi.org/10.1117/1.JBO.25.9.097002

8 Ethical Aspects in Thermal Imaging Research

8.1 INTRODUCTION

Infrared (IR) thermography is a non-invasive imaging method that measures mid-wave to long-wave IR radiation emanating from humans and converts them to temperature profiles which can be used for clinical research. Thermal imaging is done using infrared thermal cameras. They are mainly used to measure human body temperatures to find out the difference between them. They are also known as infrared thermographs or tele-thermographic system – it basically detects the infrared light emitted from a person's skin, and then it is converted into a temperature distribution map of a specific region of the body. One of the main ethical concerns is that this tool should not be used to diagnose a disease. It also should be noted that improper usage of thermal cameras can lead to health risks such as misreading the body temperature. Now they are used in airports, malls, and a lot of other public places to record the temperature. We can use it as long as you do not want it to be particularly precise. Because the temperature of a person can differ depending on the clothes that they are wearing also. Even though thermal imaging may be used for the assessment of initial temperature they are not effective when used to take temperatures of multiple people at the same time. So, it proves that thermal imaging should not be used for the screening of mass temperature. They can only measure the temperature on the surface of the skin, which is comparatively lower than the actual temperature. They will only work effectively when used in the right environment or location, always make sure that they are set up and operated correctly.

When conducting a project which uses thermal imaging various aspects have to be considered including the environment, settings in which the data are collected, the different types of information that could be captured by the camera, the different inferences that might get drawn from the images. Privacy interests are protected by the law. If the privacy of the data is mishandled, it is a punishable offense. Privacy is an idea that involves a wide range of rights and obligations meant to protect an individual from unwanted interferences into their personal domain. These advancements can encroach on people's privacy rights, especially if the individual data are subsequently imparted to aircraft, well-being specialists, or other outsiders. We are likewise worried that temperature screening is only the initial step of more meddling measures to come, like well-being or character travel papers or different kinds of well-being profiling.

DOI: 10.1201/9781003245780-8

These issues bring up significant queries for fundamental rights and assurance of people's very own information, especially if this filtering is managed without earlier consent. For as far back as 20 years, innovative headway has prompted the expanded capacity and viability of infrared imaging (IR). There have been a lot of developments in the field. Past misconceptions of IR as being lacking to viably recognize thermal profiles that are essential to building up particular differential conclusions have been soothed and invalidated by the practical refinement delivered through cutting-edge picture securing and automated logical frameworks. There is intermittent conflict encompassing the incompetent utilization of IR by the undeveloped workforce and the abuse of IR as an inappropriately controlled or deciphered "analytic" test. The previously mentioned advancement in the innovation areas of the field has extended worries over the moral utilization of IR and urges a requirement for expanded rigidity in different fields to be the only provider of this technology.

It is in this light that we examine the basics to direct and propel research that is pointed toward explaining the limits, upgrading the adequacy, and assessing the results of IR innovation as a segment of the clinical worldview of discovery, insight, and eventually finding. In any case, how this information is used is just about as significant as what this information involves, and subsequently, a brief outline of the spaces of information that keep up with the scholarly ethics significant for clinical prescription is given. If the data acquired from fundamental sciences' investigations of IR are to be the core on which its utilization in clinical practice is to be assembled, then, at that point, we should perceive the ethical commitments to use such information and specialized insight in judicious manners that help in understanding the measurements that characterize medication as a relational demonstration. Thus, the limits and impediments of IR should be perceived and elucidated to guarantee (in factual and morally) appropriate use inside the clinical fiduciary.

Hypothetical and experiential information work synergistically while relating current cases to paradigmatic models (i.e., the casuistic methodology), and thus hypothesis and experience are conjoined to decide whether, how, and why IR can and ought to be used as a stage during the time spent diagnosing specific pathology in a particular patient.

In the present circumstance, logical information permits the clinician to

1. Use specialized and experiential information on IR as an instrument in identifying the thermal marking of certain pathophysiological states.
2. Utilize this hypothesis and experience to decide if IR addresses a suitable evaluative technique.
3. Use hypothetical and experiential arrangements to detail and carry out the right conditions for the use and understanding of IR in a given case.

When using a thermal imaging system for clinical research, like the study of physiological functions, a good clinical research procedure ensures that participants' rights, safety, and well-being are protected and respected, in accordance with the principles outlined in the Declaration of Helsinki and other internationally recognized ethical guidelines, and that medical research data are reliable.

8.2   GENERAL ETHICAL PRINCIPLES

All clinical research involving human subjects should be done in conformity with the Ethical Committee's (EC) basic and general ethical principles. The researcher and the team are in charge of safeguarding the dignity, rights, safety, and well-being of the study participants. They should have the necessary qualifications and expertise in research techniques, as well as be aware of and adhere to the study proposal's scientific, medical, ethical, legal, and social requirements. The ECs are in charge of ensuring that the study is carried out in conformity with the principles mentioned below (Indian Council of Medical Research, 2017)

- **GEP1: Principle of Essentiality**: The principle of essentiality states that the involvement of human participants is required for the planned research after careful examination of all alternatives in light of existing knowledge. An ethical commission (EC) independent of the proposed research should review the research.
- **GEP2: Principle of Voluntariness**: Respect for the participant's right to consent to participate in research, or to withdraw from research at any moment, is fundamental in the principle of voluntariness. Participants' rights are protected through the informed consent process.
- **GEP3: Principle of Non-exploitation**: The principle of non-exploitation states that research volunteers should be chosen equally so that the advantages and burdens of the study are spread fairly and without discrimination. Enough measures should be in place to protect vulnerable populations.
- **GEP4: Principle of Social Responsibility**: The research is organized and carried out in such a manner that social and historic differences are not created or exacerbated, and social peace in community connections is not disturbed in any way.
- **GEP5: Principle of Ensuring Privacy and Confidentiality**: To protect the privacy of potential participants, their identities and information are kept secret, and access is restricted to only those who are permitted. However, the right to life of an individual outweighs the right to privacy of a study subject in specific instances, and this can be done in conjunction with the EC for solid scientific or legal grounds.
- **GEP6: Principle of Risk Minimization**: All stakeholders must take reasonable precautions at all stages of the research to ensure that risks are minimized and that adequate care and compensation are provided if any damage occurs.
- **GEP7: Principle of Professional Competence**: Professional competency is a principle that states that research should be planned, done, assessed, and

overseen by people who are competent and have the necessary qualifications, experience, and/or training.

- **GEP8: Principle of Maximization of Benefit**: The maximizing of benefit principle states that all efforts should be made to plan and execute research in such a way that the advantages to study participants and/or society are maximized directly or indirectly.
- **GEP9: Principle of Institutional Arrangements**: Institutional frameworks in which research institutions have rules for adequate research governance and assume responsibility for facilitating research by providing necessary facilities, staff, money, and training opportunities.
- **GEP10: Principle of Transparency and Accountability**: Principle of openness and accountability in which the study design and findings are made public through registries, reports, and scholarly and other publications while respecting the participants' right to privacy. Stakeholders engaging in research should declare and manage any existing conflicts of interest. To ensure accountability, the study should be performed in a reasonable, fair, impartial, and transparent manner. For possible external scrutiny/audit, related records, data, and notes should be kept for the appropriate duration.
- **GEP11: Principle of Totality of Responsibility**: The totality of accountability principle states that all stakeholders in a research project are accountable for their activities. All stakeholders, directly or indirectly, are bound by professional, social, and moral duties in accordance with ethical norms and relevant rules.
- **GEP12: Principle of Environmental Protection**: Environmental Protection Principle: Researchers are responsible for ensuring environmental and resource protection throughout the study process, in accordance with existing rules and laws.

8.3 GUIDELINES FOR GOOD CLINICAL RESEARCH PRACTICE

Clinical trials are an important element of clinical research since they help with the discovery of new drugs and procedures, as well as the application of existing medicines and processes. Clinical trials that are well-designed and properly conducted can assist in providing answers to important questions in health care and medication development. Their findings are critical for evidence-based decision-making. Trials that are inadequately designed and/or improperly conducted may result in a negative outcome. This jeopardizes participant safety, and evidence may be insufficient or inaccurate. They squander resources and therefore waste the investigators' and participants' efforts and time.

Clinical trials should be properly planned to protect participants' rights, safety, and well-being while also ensuring the accuracy of results. Clinical trial designs and methods must be proportional to the study's risks and the significance of the data produced. To reduce needless complexity and burden, expert committees should examine and validate trial designs and protocols.

The recommendations for good clinical practice (GCP) established by the International Council for Harmonisation (ICH) and WHO, which might be implemented while employing thermal imaging for clinical research investigations, are summarized in the following section.

The concepts outlined below provide a flexible framework for conducting clinical trials. They are designed to provide direction throughout the clinical trial's lifecycle. These concepts apply to human participants, such as healthy volunteers or patients, in clinical trials. To ensure ethical trial conduct and trustworthy outcomes, the principles are interconnected and should be evaluated in their entirety (International Council for Harmonisation (ICH), 2021)

- **GCP1**: Clinical trials should be carried out in compliance with the Declaration of Helsinki's ethical standards, which are consistent with good clinical practice (GCP) and also with any regulatory requirement(s) (WHO, 2016)
- **GCP2**: Clinical trials should be developed and carried out in such a way that participants' rights, safety, and well-being are protected.
- **GCP3**: Informed consent is a critical component of a trial's ethical conduct. Participation in clinical trials should be voluntary and based on a consent procedure that guarantees that subjects are well-informed prior to the experiment.
- **GCP4**: An institutional review board (IRB) or an independent ethical committee (IEC) should assess clinical studies objectively.
- **GCP5**: Clinical trials should be scientifically comprehensive and based on solid and current scientific information and methodologies for their intended aim.
- **GCP6**: Clinical trials should be planned and carried out by professionals.
- **GCP7**: Clinical trials should be designed and conducted with quality in mind, both scientifically and operationally.
- **GCP8**: Processes, methods, and procedures in clinical trials should be appropriate to the risks to participants and the dependability of trial outcomes.
- **GCP9**: Clinical trials must be specified in a protocol that is clear, simple, and operationally realistic.
- **GCP10**: Clinical studies should produce consistent outcomes.
- **GCP11**: Clinical trial roles, duties, and obligations should be well defined and recorded.
- **GCP12**: Investigational items such as thermal imagers used in clinical trials should be prepared in compliance with good manufacturing practice (GMP) standards and stored, delivered, and handled in line with the product specifications and trial protocol.

WHO Principles of GCP are explained in the following section (WHO GCP 2002):

- **WHO1**: Human research should be based on good science and carried out in conformity with basic ethical standards outlined in the Declaration of Helsinki. All other GCP values are pervaded by three core ethical principles of equal importance: respect for individuals, beneficence, and fairness.
- **WHO2**: Human research should be justified scientifically and presented in a clear, thorough methodology.
- **WHO3**: Prior to beginning human research, all potential hazards and discomforts, as well as any expected benefits to the individual trial participant and society, should be identified. Non-clinical and, where applicable, clinical data should be used to support research on experimental items or treatments.
- **WHO4**: Human research should only be undertaken if the expected benefits to the individual study subject and society clearly exceed the dangers. Although the benefits of the trial's results to technology and society should be considered, the rights, safety, and well-being of the research participants are the most critical concerns.
- **WHO5**: Prior to beginning human research, an independent ethics committee/institutional review board (IEC/IRB) approval/positive opinion should be obtained.
- **WHO6**: Human research should be carried out in accordance with the authorized procedure.
- **WHO7**: Prior to study involvement, every subject should offer freely given informed permission in line with country culture(s) and standards. When a person is unable to give informed consent, a legally authorized representative's approval should be sought in line with applicable legislation.
- **WHO8**: Human research should only be continued if the benefit-risk profile remains favorable.
- **WHO9**: Medical staff who are qualified and officially licensed (i.e., a physician) should be in charge of the study subjects' medical care and any medical decisions made on their behalf.
- **WHO10**: Each person engaged in a trial should be qualified to execute his or her separate task(s) based on education, training, and experience, as well as be currently licensed to do so, if needed.
- **WHO11**: All clinical trial data should be captured, processed, and maintained in such a way that it can be accurately reported, interpreted, and verified.
- **WHO12**: The confidentiality of documents that potentially identify individuals should be safeguarded, in line with existing statutory requirements for privacy and confidentiality.
- **WHO13**: Investigational goods should be made, handled, and stored in line with GMP and utilized according to the authorized procedure.
- **WHO14**: It is necessary to develop systems and processes that ensure the quality of every component of the trial.

8.4 RESPONSIBLE CONDUCT OF RESEARCH

The value and rewards of a study are determined by the researchers' honesty. Scientists bear a huge social obligation to protect the public against research malpractice and misuse. Responsible researchers follow the rules set out by their professions, specialties, and institutions, as well as applicable laws. A research team's members are expected to maintain high standards and support research's essential ideals. The following key components of responsible conduct of research (RCR) are involved: values; policies; planning and doing research; evaluating and reporting research; and responsible authoring and publishing of clinical research outcome (Indian Council of Medical Research, 2017).

Shared values such as morality, precision, efficiency, equality, impartiality, reliability, accountability, openness, personal integrity, and awareness of current best practices lead RCR, and these values should be represented in RCR rules. Scientific research is critical for bettering our knowledge of multiple health issues and their remedies. As part of inter-professional ethics, all research components rely on collaboration and shared expectations. Immoral practices in scientific research have the potential to erode public faith in science while also having a detrimental influence on the research team.

Research participants must be protected by rules and methods established by institutions. The institution, the Ethics Committee, and the researchers should all be assigned obligations under such policies. There should also be systems and rules in place to oversee research, such as data acquisition, management, conflicts of interest, scientific misconduct reporting, and adequate initial and ongoing training for researchers and EC members. The policies can be found on the institutions' or organizations' websites. Researchers should also adhere to their professional codes of ethics.

8.5 ETHICAL ISSUES REVIEW PROCEDURE

To protect the dignity, rights, safety, and well-being of all research participants, all study proposals in biomedical, social, and behavioral science involving human participants, their biological material, and data must be evaluated and authorized by a suitably constituted EC. ECs are responsible for the initial examination of research studies prior to their implementation, as well as the ongoing monitoring of authorized research to guarantee ethical compliance. In order to function properly, the EC must be competent and autonomous. The organization is responsible for developing an EC to guarantee a competent ethical review and monitoring system that is suitable and sustainable. Infrastructure, personnel, space, funding, proper assistance, and secured time for the Member Secretary to conduct the EC activities are all responsibilities of the organization. The EC is in charge of reviewing research proposals for ethical and scientific reasons.

Although ECs may use evidence from a previous systematic study, they must ensure that the study methodologies are scientifically sound and that the chosen research plan or approach has ethical consequences. Before being done, all sorts of biomedical and health research (clinical, fundamental scientific, policy, implementation, epidemiological, psychological, public health research, and so on) must be evaluated by an EC. Multidisciplinary and multi-sectoral ECs are desirable. There should be proper age and gender representation. Non-affiliated or external members should make up at least half of the membership. An EC's size should ideally be between seven and fifteen members, with a minimum of five members present to fulfill quorum requirements. Depending on the needs of the institution, the EC should have a balance of clinical and non-clinical members/ technical and non-technical members. The EC can establish a panel of topic experts who are sought for their subject knowledge, such as a pediatrician for child research, a cardiologist for heart disease research, and so on. They may be asked to the meeting to provide an informed advice on a specific proposal, but they will not be able to vote or make decisions. Before referring a proposal to the EC, a special scientific committee should evaluate it as much as feasible. EC can pose scientific questions in addition to ethical ones, as both excellent science and ethics are required to assure the quality of experiment and the safety of participants.

8.6 INFORMED CONSENT PROCESS

For any biological or health study involving human subjects, the researcher must get voluntary written informed permission from the prospective participant. This condition is founded on the idea that competent persons have the right to freely choose whether or not to engage in the study. Informed consent is a continual process that includes three key components: presenting relevant information to potential subjects, guaranteeing the individual's competency, ensuring that the material is easily understood by the participants, and ensuring the voluntariness of participation. Informed voluntary consent preserves and respects an individual's freedom of choice and autonomy. After going through the informed consent process of obtaining information, comprehending it, and willingly deciding to engage in the research, each candidate shall sign the informed consent form.

8.7 ISSUES RELATED TO MEDICAL DATASETS

With the ease with which large repositories can be set up and maintained, the primary goal of data storage and processing in some of these datasets may not be research, but with advancements in information technology (IT) and lower costs, they offer a huge potential for future research and value addition. When such archives are utilized for study or future commercialization, they must adhere to the same standards as any other health-related research. Outside of typical healthcare settings, health records are increasingly being gathered.

Information is shared with other parties for a variety of reasons, including research and commercial benefit. Big data in biomedical research raises a wide range of ethical considerations, ranging from potential threats to personal rights, such as privacy, to questions about individual autonomy. There are other features that are unique, such as the data sources, scalability, and open access policies. The ethical problems surrounding big data, such as information security, sharing, rights, benefit-sharing, and others, must be thoroughly investigated.

8.8 GENERAL GUIDELINES WHEN USING THERMAL IMAGING FOR CLINICAL SCREENING

For many illnesses, early detection and treatment are critical. It enables the use of more preemptive and less intrusive treatment approaches. Early illness and injury identification also provide for a quicker recovery. Thermal imaging may also be used to locate and highlight areas of the body that are experiencing inexplicable pain or injuries. This might lead to a faster and more accurate detection of pain and trauma areas that other screening devices could overlook. Thermal imaging is safe since it does not utilize ionizing radiation and has no negative effects. It is appropriate for people of all ages, including small children and newborns. Thermal imaging is a complementary test that aids in the differential diagnosis process but is not diagnostic of disease on its own. Digital infrared thermal imaging has a wide range of medicinal uses, including rheumatology, neurology, oncology, physiotherapy, and sports medicine. Thermal imaging systems are a low-cost, easy-to-use tool for swiftly and precisely assessing and monitoring patients.

For optimal thermal data capture, we must prepare the area where you will use a thermal imaging system. The room temperature should be between 68°F and 76°F (20°C and 24°C), with a relative humidity of 10–50%. To reduce reflected infrared radiation, avoid reflective backdrops (such as glass, mirrors, and metallic surfaces). Use in a room that has no draught (air movement), is not in direct sunlight, and is not exposed to radiant heat (e.g., portable heaters, electrical sources) (Ring & Ammer, 2012). Strong illumination should be avoided (e.g., incandescent, halogen, and quartz tungsten halogen light bulbs) (Center for Devices, and Radiological Health, 2021). To ensure that readings are correct, certain systems call for the use of a calibrated blackbody (a tool for testing the calibration of an infrared temperature sensor) during assessment. To see if a calibrated blackbody is required, consult the manufacturer's instructions. Some gadgets don't need one at all. Warm up the system by turning it on 30 minutes ahead of time.

Preparing the subjects before being measured is crucial for proper thermal imaging. It should be noted that there are no obstacles to the face, such as a hat, scarf, spectacles, or face shield. Hair should be pushed back from the person's face, and the person's face should be clean and dry. Moreover, according to an FDA report, the benefits of wearing a mask over the mouth and nose while utilizing thermal imaging devices exceed any potential risk of erroneous readings during the COVID-19 health emergency. Wearing excessive clothes or head covers (e.g., headbands, bandanas) or using facial washing solutions does not result in a greater or lower face temperature (e.g., cosmetic wipes). It should be noted that the subject has been in the measuring room for at least 15 minutes or 30 minutes after exercising, severe physical activity, bathing, or applying hot or cold compresses to the face (Ghassemi et al. 2018).

8.9   ETHICAL CONSIDERATIONS IN THERMAL IMAGING RESEARCH

The temperature of our bodies can reveal a great deal about our health. Different types of temperature measurements are inextricably linked, particularly when infrared thermal imaging is used to monitor human body temperature. The normal skin temperature is 33°C (91°F). The flow of energy to and from the skin determines the sensation of heat. Thermography is a technique for determining the distribution of heat. In medical imaging, thermography is used to determine the body's surface temperature. Thermal pictures can be derived from a blackbody, an ideal object that absorbs electromagnetic radiation, i.e., light, and whose radiation absorption is proportional to its radiation emission. Because human skin has a high emissivity of 0.98 out of 1.0, it is possible to establish a link between temperature and light emitted by the skin. A skin temperature of approximately 33°C, or 306 K, emits light with a wavelength of 9.5 µm, referred to as long-wavelength infrared radiation. A rise in body temperature is a natural sign of disease.

Infrared imaging makes use of specialized and powerful cameras called infrared imagers to detect the electromagnetic spectrum's long infrared range. Between 8 and 15 m, a long-wave infrared sensor gives thermal data or a heat map of the captured scene in a quantified format that represents temperature. Because thermography is non-contact and non-invasive, it is a patient-friendly method. Thermal cameras are significantly more expensive than visual cameras.

When used as a medical instrument, thermal imaging can be a very useful screening tool for determining the temperature of the human body in relation to suspected infection or sickness. This technique is convenient and efficient, and it may be used in a touchless/non-invasive manner to identify candidates for further diagnostic testing in a matter of seconds. The use of this screening approach may help to prevent and slow the transmission of contagious disease, as it allows the operator to operate the instrument at a wider distance from the patient than forehead-reading thermometers allow. Thermal imaging is a two-step procedure in which cameras capture and measure the infrared radiation generated by a person and then link it to a visible display for real-time analysis. Pattern recognition software visualizes changes in the skin's surface temperature as they occur. Through the use of a monitor, a digital colorized heat map of the face can be easily analyzed to indicate the presence of a suspected underlying health issue.

8.9.1   INFRARED IMAGING SYSTEMS AS MEDICAL DEVICES

Thermal imaging devices are being employed as a front-line tool for measuring human body temperature and recognizing those with underlying medical issues during this time of global health crisis. Despite this, not all thermal imaging systems in use today exceed regulatory criteria or medical device specifications. This raises the danger of making an incorrect diagnosis when non-medical thermal imaging is used to diagnose health problems.

Thermal imaging is also well-suited for a range of healthcare applications due to the fact that the amount of infrared radiation emitted by an object rises with temperature. Thermography is currently most frequently employed in oncology, neurology, and rheumatology as a diagnostic tool, and is usually recognized as the preferred way for diagnosing breast diseases, certain types of blood vessel difficulties, arthritis, neuromuscular conditions, and other forms of nerve damage.

Thermal imaging systems and gadgets for medical applications have a number of major advantages over conventional diagnostic equipment. To begin, thermal imaging is a non-invasive method that requires no surgical incisions and provides no potential safety threats to patients. Additionally, thermography devices can capture the data necessary for an effective diagnosis without contacting the patient directly.

Simultaneously, because of their high sensitivity, thermography systems and devices used in medical diagnosis are often utilized in highly controlled situations to assist verify the validity of testing results and to minimize the occurrence of false-positive or false-negative findings. Temperature, humidity, draughts, or other excessive air movements, and direct sunlight or intense lighting can all have an effect on the reliability of thermographic technology in diagnostic procedures.

Other aspects that may affect the usefulness of thermography systems in medical diagnostic applications include the training provided to users on how to properly use the technology and whether patients have been appropriately prepared for thermographic examination. For example, in the latter situation, strong physical activity or exercise soon prior to testing can cause temperature readings to rise, potentially falsifying the test results.

Thermal pictures can assist doctors in diagnosing and managing patients by determining whether the condition is a functional problem, its location, and its severity. Additionally, the process may aid the practitioner in assessing the patient and determining the most effective course of action. Thermal imaging is a legitimate analytical technique that can be conducted by a medical practitioner who has been trained and certified by a reputable organization. Thermal pictures should, however, be analyzed by licensed clinical specialists or clinical thermographers. The purpose is to ensure that if the images are accurately interpreted, the patient receives targeted care and treatment recommendations.

8.9.1.1   Standards of Thermal Imaging System

Medical electrical equipment – Part 1, IEC 80601-2-59: General requirements for basic safety and essential performance which specifies thermographic evaluation requirements not covered by IEC 60601-1. IEC 80601-2-59 is a standard that is part of the IEC 60601 series of standards for medical electrical equipment. It was last updated in 2017.

Medical electrical equipment – Part 2, IEC 80601-2-59: The internationally recognized standard that outlines specific safety and function requirements for thermal imaging systems used as medical devices for human temperature screenings is basic safety and essential performance requirements for screening thermographs of human febrile temperature. (Center for Devices, and Radiological Health, 2021)

Following are some of the key terms and concepts enclosed in this standard:

- **Target Area**: The required target measurement site of a thermal imaging device is defined by the standard as the median area between the inner corners of each eye.
- **Temperature Imaging Range**: The device must be capable of producing a thermograph of the face at temperatures ranging from 30°C to 40°C.
- **Temperature Accuracy**: The laboratory accuracy of temperatures measured by the device must be less than 0.5°C minus the system's determined ambiguity over the 34–39°C range.
- **Threshold Temperature**: The device's alarm threshold temperature must be adjustable in increments of no more than 0.1°C and must cover a temperature range of at least 34–39°C.
- **External Temperature Reference Source**: The standard specifies the size and performance requirements for the device's mandatory external temperature reference source, which is typically a blackbody radiator in the camera's field of view at a setpoint for useful comparisons to human body temperature.
- **Drift and Stability Measurement**: The drift and stability of the device must be less than 0.1°C over a 14-day period or the calibration interval stated for the device, which will be much longer.
- **Minimum Resolvable Temperature Difference**: The device's minimum resolvable temperature difference (MRTD) must be less than 0.1°C. To ensure compliance with this requirement, the standard refers to the testing procedure in ASTM E1213-14.
- **Uniformity of Workable Target Plane**: The maximum temperature difference that can be measured at different locations across the target area is 0.2°C. The standard specifies the test methodology that will be used to determine compliance with this requirement.
- **Face Position**: The functional target area of the device must be able to screen a face positioned between 0.75 and 2.2 m above the floor. The plane of the scanning device must be parallel to and in line with the face.
- **Spatial Resolution**: At the workable target plane, the horizontal and vertical spatial resolution of an image pixel generated by the system must be less than or equal to 1 mm. The resolution across the typical human face is 240 × 180 mm, imaged by at least 240 × 180 image pixels.

Other important requirements detailed in IEC 80601-2-59 for thermography systems used for human temperature screening include the following:

- **Alarm Conditions for Screening Thermography (Clause 201.102)**: A physiological alarm mechanism must be included in the thermal scanning device to indicate when the temperature recorded within a target area exceeds the prescribed threshold temperature. A separate alarm mechanism must also be included in the device to prevent a newly activated device from being used before it has completed the device start-up process.
- **Control and Instrument Accuracy, and Protection against Hazardous Outputs (Clause 201.12)**: In addition to meeting the relevant requirements in Clause 12 of IEC 60601-1, the thermal scanning device must also meet new requirements in IEC 80601-2-59 for device display. The standard addresses specific aspects such as display size, color scale, temperature resolution, response time, and throughput.
- **Identification, Marking, and Documentation of Equipment (Clause 201.7)**: The standard refers to the requirements for identification, marking, and documentation in Clause 7 of IEC 60601-1, but also requires additional operating instructions to ensure that the face being scanned is not obstructed by hair, eyeglasses, or other objects. When the thermal scanning device detects a temperature reading within the target area that exceeds the prescribed threshold temperature, IEC 80601-2-59 also recommends that the operator perform a secondary screening with a clinical thermometer.

8.9.1.2   Handling Thermal Imaging Systems

It is crucial that the user who handles the device follows all manufacturer instructions to determine whether the system is properly set up and located so that it can precisely measure skin temperature. It is important that the operating person is trained to prepare the location and the participant being evaluated in order to improve the accuracy.

(a) **Proper Thermal Imaging Room Setup**:
- The temperature in the room should be between 68°F and 76°F (20°C and 24°C), with a relative humidity of 10–50%.
- Other relevant factors that may influence temperature measurement should be kept under control. They are explained as follows:
  - Avoid reflective backgrounds to reduce the reflected infrared radiation.
  - It is best to use the device in a room that is free from drafts, away from direct sunlight, and away from radiant heat sources.
  - Avoid using bright lighting.

(b) **Environmental Control**:
Ambient temperature is modified so that the physiological state of the patient does not significantly cause shivering or excessive sweating. It should range from 18°C to 23°C. To maintain the stability and permit all the areas of the body to acclimate equitably, the thermal condition of the room must be gradual during an examination. The ambient temperature should not differ by more than 1°C during the study. Also, the humidity of the room should be controlled so that no air moisture, perspiration, or vapor levels interact with radiant infrared energy. A room thermometer is required to effectively monitor the temperature of the image acquisition site.

(c) **Proper Thermal imaging System Setup**:
 • During the evaluation process, the accuracy of measurements is inspected by using a blackbody calibration.
 • Allow the entire system to warm up for 30 minutes before use.

8.9.2 PATIENT MANAGEMENT PROTOCOL

Thermal artifacts can be reduced and image accuracy improved with proper patient management, both before and during an exam.

8.9.2.1 The Pre-Examination Process

To reduce thermal artifacts, pre-examination preparation guidelines are required. Prior to the examination, the patient should be given the following minimally necessary instructions:

 • There should be no lotion, cream, powder, or makeup applied to the body part being imaged.
 • No deodorant or antiperspirant may be used during the upper body imaging procedure.
 • Avoid physical therapy, physical stimulation, or use of hot or cold packs for at least 24 hours before the examination.
 • Exercises should be avoided four hours before the exam.
 • Should take bath at least 1 hour before the exam.
 • Avoid taking pain relievers and drugs unless it is advised by the patient's doctor. Before changing any medications, the patient should consult with their doctor.

(a) **Patient Acclimatization:** The patient should be provided enough time to adapt to the ambient temperature so that the thermodynamic equilibrium could be achieved. Allow at least 15 minutes for equilibration. Additional equilibration time results in minimal surface temperature changes. The region being imaged should be free of clothing and accessories during the equili-bration phase and subsequent assessment. During the equilibration phase, a loose-fitting robe may be worn to provide modesty prior to some procedures, as long as it does not limit the airflow or stiffen the skin surface in any form that might provide an inconsistent thermogram.

(b) **Imaging Protocol:** Intensive laboratory environmental standards and patient preparation recommendations ensure that the subject is physiologi-cally prepared for thermal imaging.

   i. **Patient Positioning:** For thermal imaging, both the electronic infrared and liquid crystal thermographic systems require fundamental standard patient and equipment positioning. In order to study the physiology of these loca-tions, the entire upper or lower body is routinely scanned. As needed, spe-cialized or limited perceptions can be added or examined individually.

   ii. **Using a Thermal Imaging System:**
- During each measurement, only one person's surface skin tempera-ture should be measured.
- The person should be positioned in front of the thermal imaging system at a fixed distance.
- Ideally, the person's entire face and a calibrated blackbody should be included in the image area.

(c) **Data Documentation:** Each thermal image captured should include an indication of anatomical information and also the following details should be included with the original image or be easily accessible to other archived documents:
1. Name of the patient/Identification number
2. Date on which the image was captured
3. The name and address of the imaging facility

8.9.3 INTERPRETATION AND REPORTING

The analysis of thermal data includes every aspect of thermal imaging and its rel-evance to human physiological processes and systems. Sensory and sympathetic nerve systems, circulatory system, musculoskeletal system, endocrine system, as well as the localized inflammatory processes are all investigated during the inter-pretation process. A clinical impression is made by combining thermal informa-tion of the image with clinical data.

(a) **Reporting Format:**
The following information should be included at a minimum in the report-ing format:
1. Address of the imaging facility
2. Name of the patient
3. Age of the patient

4. Date of examination
5. Clinical data
6. Symptomatology
7. Significant thermal imaging findings
8. Clinical impression
9. Suggestions
10. Qualified interpreter's signature

(b) **Guidelines for Clinical Thermography Education:**
Adequate thermal infrared imaging training is necessary to ensure proper data acquisition, data analysis, and public safety.

(i) **Certified Clinical Thermal Imaging Technician:**
Certification courses comprise both theoretical classroom hours and also practical imaging experience. Principles of thermography, patient management, laboratory and imaging protocols, and supervised practical field experience are generally covered in the training. Training on image interpretation is not provided to the technician. Certified clinical thermographic technologists are those who have successfully completed the relevant course.

(ii) **Certified Clinical Thermologist:**
This level of training combines formal classroom instruction with hands-on experience. The course material involves discussion of appropriate physiology and anatomy, disease pathogenesis mechanisms and their connection to thermal imaging lectures, imaging protocols, patient management, thermal imaging fundamentals, image interpretation, and analysis. Board-certified clinical thermologists have earned their credentials after successful completion of the course.

(iii) **Breast Thermologist:**
Extended breast thermography training is highly suggested for doctors who intend to interpret thermographic images of the breast. This level of training often surpasses the course content leading to certification in thermographic breast imaging.

8.10   SUMMARY

Thermal imaging is a non-invasive imaging technique that uses a thermal sensor to detect heat radiations emitted by the study subject in order to determine the temperature profile of the subject's body surface. Even if non-invasive technology such as thermal imagers are used, the experimenter/investigator of any medical study should ensure the safety of the patients and the security of their medical data. The relationship between the experimenter/doctor and the patient is one of mutual trust and respect. The investigator's main goal is to serve humanity while maintaining the dignity of his or her profession and of man.

APPENDIX I

Proforma to Be Submitted to the Institute Ethics Committee (Human Studies) (template)

1. Project title:

2. Principal investigator/co-investigators with designation & department:

3. Sources of funding:

4. Primary & secondary objectives of the study:
   - Primary objective:
   - Secondary objective:

5. Background & justification for the conduct of the study:
6. Study hypothesis/research question:
7. Methodology:

   a. Sample size with justification
   b. No of groups:
   c. Inclusion criteria:
   d. Exclusion criteria:
   e. Intervention:
   f. Control:
   g. Dosages of drug & frequency with duration:
   h. Investigations/procedures to be done etc.:
   i. Type of randomization & method used:
   j. Method of allocation concealment:
   k. Blinding/masking if any:
   l. Brief procedure:

8. Setting in which subjects will be recruited from:
9. Period of recruitment:
10. Potential risks involved to the participants of the study:

11. List out the benefits expected by the participant as a result of study participation:
12. Do you need exemption from obtaining informed consent from study subjects – if so, give justification:

13. Whether consent forms I and II in English and in native language are enclosed?

14. If appropriate, is there a children's assent? If yes, please submit a copy of this form:
15. Documents attached for regulatory clinical trials:
16. Conflict of interest for any other investigator(s) (if yes, please explain in brief):

Signature of the Investigator ——————Date:
Signature of the Head of the Department ——————Date:

Note: Consent forms I and II in English and the patient's native language must be included with the pro forma. The patient information sheet is the same as a consent form. The investigator must communicate with the subjects in plain English and in a dialogue manner.

APPENDIX II

INFORMED CONSENT FORM (TEMPLATE)

Study title:

Study number:

Subject initials: Age:

Subject name:

I certify that I have read and comprehended the above-mentioned information document. I got the opportunity to ask questions, and all of them were well addressed. I realize that my participation in the study is entirely voluntary and that I am free to withdraw at any time for any reason, without affecting my medical care or legal rights. Even if I withdraw from the trial, I understand that the investigator and the ethics committee will not require my consent to look at my health information for the present study and any future research that may be undertaken in relation to it. This access is OK to me. However, I realize that any information shared with other parties or publicized will not expose my name. I agree not to limit the use of any data or results generated by this study, as long as they are used solely for scientific purposes.

I agree to take part in the above study.

Signature (or thumb impression) of the subject:—————Date:

Signatory name:—————Signature of the legally

accepted representative

—————Date:—————Signatory name:—————

Signature of the investigator—————Date

Study investigator name:—————

Signature of the witness:—————

Date:—————

Name of the witness:—————

REFERENCES

Center for Devices, and Radiological Health. 2021. Thermal imaging (infrared thermographic systems/thermal cameras). U.S. Food and Drug Administration. Retrieved March 5, 2021, from https://www.fda.gov/medical-devices/general-hospital-devices-and-supplies/thermal-imaging-systems-infrared-thermographic-systems-thermal-imaging-cameras

Center for Devices, and Radiological Health. n.d. "Medical Devices." U.S. Food and Drug Administration. FDA. Accessed July 6, 2022. https://www.fda.gov/medical-devices.

Ghassemi P, Pfefer TJ, Casamento JP, Simpson R, Wang Q. Best practices for standardized performance testing of infrared thermographs intended for fever screening. *PLoS One* 2018, 13(9): e0203302. https://doi.org/10.1371/journal.pone.0203302

Indian Council of Medical Research. 2017. National ethical guidelines for biomedical and health research involving human participants. Retrieved January 10, 2022, from https://ethics.ncdirindia.org//asset/pdf/ICMR_National_Ethical_Guidelines.pdf

International Council for Harmonisation (ICH). 2021. Ich e6 guideline for good clinical practice (gcp)-update on progress public web conference report. Retrieved January 10, 2022. from https://database.ich.org/sites/default/files/ICH_E6R3_WebConference_Report_Final_2021_1011.pdf

Ring EF, Ammer K. Infrared thermal imaging in medicine. *Physiological Measurement* 2012, 33(3): R33–R46. https://doi.org/10.1088/0967-3334/33/3/R33

WHO. 2016. Handbook for good clinical research practice (gcp). Retrieved January 10, 2022, from https://apps.who.int/iris/handle/10665/43392

Index

A

abdomen, 18, 37, 160
abnormal, 47–48, 121, 127, 134–135, 143–144
abnormality, 25, 127, 166, 184
acclimation, 31, 37–39, 98, 126, 189
active contour, 18, 69–70
AdaBoost, 83, 194–196
Adadelta, 89
AdaGrad, 89
Adam, 89, 169
adaptive moment estimation, 89
adjuvant-induced, 97, 102
affine transformation, 15, 61
agglomeration, 106, 115
amalgamated, 13, 100
ambient temperature, 31, 33–36, 38, 40, 108, 128, 158, 182, 187, 189, 218
amputation, 159–160, 166
anthropometrical, 159
antigen-induced, 106
anti-inflammatory, 16, 20
arrhythmia, 8
arthritis-induced, 106, 109–110, 119
asymmetric, 158, 169, 178
automated thresholding, 68, 131–132, 143

B

backpropagation, 88–89
batch normalization, 87, 92–93
β-cells, 155
bilateral, 22, 101, 127
binarization, 9
Biomarkers, 41–42
bootstrap, 83
bounding boxes, 8
breathing patterns, 5, 8
breathing rate, 5–7, 45

C

calcaneal, 170–171
calculus, 56
Canny edge detection, 72, 134
cardiac, 9–10, 148
cardiovascular, 2, 8
cartilage-specific, 106
classifications, 81, 83, 90–91, 176
clinical disease activity index, 124

clinical infrared radiometers, 3
clustering-based, 64, 144
coalitional tracking algorithm, 6
collagen induced arthritis model, 113
complex regional pain syndrome, 2, 21–22
computer-aided, 17, 63, 169, 173
constant-temperature-controlled, 107
contrast enhancement, 10, 58
convolutional, 73, 91–92, 198
Convolutional Neural Networks, 73
counterstream mechanism, 1
cross-correlation, 13
cross-entropy, 199
cross-validation, 140, 142, 144

D

decision level fusion, 16
decision-making, 92, 208
decomposition, 62–63, 80
deep-learning-based, 159, 172, 176
deep neural network, 84
dendrogram, 65
density-based clustering, 66
dexamethasone, 98, 105, 113
diabetes, 2, 18, 122, 148, 155–159, 161, 164, 166, 168–169, 173–174, 176–178
diabetes complications, 2, 18, 173, 178
diabetics, 48, 178
diagnose, 20, 25, 156, 205, 215
diffusion, 4, 9
dilation, 2, 8, 16, 33, 74, 136
directional emissivity, 29
disability, 123, 145, 178
discontinuities, 70, 132, 134, 144
discriminant, 10, 140–141, 175
disease activity index, 122, 124
disorder, 14, 21, 144, 155, 177, 190
dissected, 105, 112, 117
dissimilar, 64, 69
distal–dorsal, 46
distal interphalangeal joint, 17
divide-and-conquer, 141
divisive analysis clustering, 65
divisive clustering, 64–65
double thresholding, 73
downsampling, 86, 93
dropout, 176
dystrophy, 21, 23

E

edge detection, 56, 64, 70–72
efficacy, 100, 143, 168, 209
eigen, 11
electronic thermometer, 34
encephalography, 12
entropy, 75, 78, 84, 174
erosion, 8, 16, 103, 117, 136
exploratory, 169

F

facial recognition, 15
facial thermal image, 6, 14, 20
facial thermograms, 158–159, 161–162, 177,
        184, 189, 192–193
facial tracking, 5
factorization, 93
false-negative, 19, 24, 190, 215
false-positive, 19, 24, 215
fast Fourier transform, 4
feature-based, 174
feature descriptors, 138, 174–175, 197–198
feature extraction, 14–15, 56, 62, 66, 73–76, 80,
        87, 121, 138, 144, 174, 176, 195, 197
feature matching, 75, 138
feature points, 10, 62, 138, 174–175
feedforward, 35, 87, 177
fever-screening, 192–193, 200
fuzziness, 67
fuzzy means, 17

G

Gaussian, 8, 58, 72, 75, 79, 175
gestational, 155
Glamorgan, 30, 44–45, 47, 160
global thresholding, 131
glucose, 155–160, 162–164, 166, 177
gradients, 19, 21, 46, 48, 88–89, 131, 197–198
granular, 20, 58
graph partitioning, 64
gray level co-occurrence matrix, 17, 74–75,
        78, 174
gray level run length matrix, 74
gray level size zone matrix, 74, 77
grayscale, 57–58, 67, 70, 73, 77, 79, 131–132,
        134, 136–137, 146, 164
guideline, 2

H

Haar-cascades, 197
hamming, 138
handcrafted, 121

hazardous, 217
health assessment questionnaire, 122
healthcare, 212, 215
heart rate detection, 9
heat capacity, 3
heat distribution index, 124–125
heat-sensing, 191
heat transfer coefficient, 3
hemoglobin, 125, 159, 164
hierarchical clustering, 64–66
higher-order, 74, 169
high-grade, 47
high-temperature, 8
histogram, 12, 33, 48, 59–60, 64, 68–69, 77,
        131, 197–198
histogram equalization, 59–60
histopathological, 106, 112, 114–115, 117–119
homeotherms, 4
homogeneity, 75, 174
homography, 62
hotspot, 134–136, 144
humidity, 31, 35–36, 39, 99, 104, 122, 161, 213,
        215, 217
hyperglycemia, 155–158
hyperparameter, 89, 142
hyper-planes, 82
hypertension, 8
hyperthermia, 17–18, 23, 38, 182
hyperthermic joints, 17
hypodermic, 101–102
hypothermia, 38, 182

I

idiopathic, 121
illness, 25, 181, 213
illumination, 108, 213
image acquisition, 107–108, 122, 126, 160,
        184, 218
image enhancement, 56, 58
image fusion, 61–63
ImageNet, 91
image segmentation, 121, 128, 130, 136, 143
image sharpening, 58–59
image smoothening, 58
immunization, 98–99, 105, 110
impediments, 159, 206
impression, 117, 219, 222
incandescent, 161, 213
incidence, 100, 113, 166
inductance plethysmography, 5
inexpensive, 25, 32, 158
infections, 157, 160, 168–169, 178
inflammation, 16–17, 19, 25, 97, 99–100, 102,
        104, 107–108, 112–113, 117, 121,
        127, 145, 166, 168, 177

inflammatory arthritis, 1, 16, 97
infrared-based, 3, 30, 189
infrared thermography, 1–3, 18–21, 30, 40, 158, 178
injuries, 21, 23, 42, 213
intelligence-based, 192
intensity-based, 61, 144
intercostal, 107
interpolation, 15, 70
intervention, 41, 221
intrusive, 213
invariance, 138
investigation, 14, 19, 23, 41–42, 111, 146, 151, 166, 190
ionizing, 158, 213
ipsilateral, 168
ironbow, 57–58
ischemia, 169
isotherms, 48–49

J

joint, 16–18, 43, 97, 100–105, 107, 112, 114–115, 117–119, 121–126, 128, 130, 145–150
juvenile, 121, 155
juxta-articular, 122

K

keypoint, 75, 79, 138
k-means, 17–18, 64–66, 130, 135, 144
k-nearest, 15, 80–81, 140, 175
K-star, 80, 83–84

L

landmarks, 43, 47, 193, 199–200
Laplacian, 58–59, 72
laser, 46, 125
latencies, 103–104
lateral, 18–21, 34, 40, 44, 125, 146, 170–171
lesion, 20, 70
ligaments, 118
linear discriminant analysis, 140, 175
lipopolysaccharides, 98, 184
local binary gray level co-occurrence matrix, 74
local binary pattern, 74, 77–79
local directional pattern, 74
localization, 11, 72, 79
locomotor, 1–2, 25
log edge detection, 134
logistic, 80, 83
LogitBoost, 80, 83
long-wave, 205, 214

losses, 89
lymphocytes, 104, 112, 114–116, 184

M

machine-learning, 56, 159
machine-learning-based, 168, 174
malignancy, 19
Mann–Whitney, 98, 105
manual thresholding, 131–132, 143
Marr-Hildreth, 72
Mask-R-CNN, 74
mass-screen, 24, 200
maximization, 208
max-pooling, 86, 91–92, 199, 201
mean shift clustering, 64, 67
mean-squared, 199
medial, 12, 18, 20, 101, 146, 170–171
median, 22, 112, 141, 216
metacarpophalangeal, 16–17, 124
metacarpophalangeal joints, 16
methodology, 168, 206, 210, 216, 221
microbolometer, 16, 108, 122
microcirculation, 5
microvascular, 4
midway, 198
misalignment, 61
MobileNet, 93
molecular, 182, 184
momentum, 75
mono-arthritis, 100, 113, 117
monoarthritis model, 100
monochrome, 3
morphological, 8, 11, 16, 18, 73–74, 136, 174
morphological operations, 8
morphology-based, 11, 137
mucosa, 20
multi-cascade, 200
multi-class, 140
multilayer, 199
multi-level thresholding algorithm, 10
multispectral, 60–62, 93
multispectral images, 60–62, 93
musculoskeletal, 219
myofascial, 49

N

naïve, 80–81, 142–144, 175
nearest, 65, 75, 101, 138–140
near-infrared, 111
neighborhood, 58, 66, 68, 75, 174, 198
neutrosophic, 64, 67
non-invasive, 1, 9, 21, 25, 97, 121, 145, 157, 159, 205, 214–215, 220

non-linear, 79, 87, 92
non-linearity, 33, 86
non-maximum suppression, 73
non-uniformity, 32, 76, 78
normoglycaemia, 157–158
Normothermia, 182

O

obesity, 156, 166, 169
object detection, 58, 62, 73, 91, 174
open-source, 174
optimizer, 87, 89
Optotherm, 24, 187
ORB-based, 174
oriented FAST feature point selection, 174
orofacial, 2
osteoarthritis, 1, 121, 125
outliers, 66, 128
over-segmentation, 144

P

pain-related, 101–102
pandemic, 181, 186
paraffin, 106, 112, 117–118
parameter, 2, 15, 21, 61, 75, 77, 79, 82, 89,
        101, 151
partition, 18, 63, 68, 107
patellofemoral, 121
pathogenesis, 97, 220
pathological, 97, 112, 118, 121
patterns, 5, 8, 14–15, 19, 25, 75, 78, 127,
        158–159, 166, 184
perception, 2, 60, 62
perfusion, 5, 21, 146–151
perinasal region, 11
periodicity, 11
peripheral skin temperature, 38
perspiration, 4–5, 11, 218
perspiratory responses, 11
pharmacological, 107, 121
phase-contrast, 102
physiology-regulatory, 161
physiotherapy, 31, 42, 122, 213
pixel-based, 73, 194
planar, 62
plantar, 13, 18, 103–104, 159, 166, 169–172,
        176
plasma, 112, 114–115, 156–158
plethysmograph, 5, 99
point-based, 61
polyarthritis, 106, 113
pooling, 85–86, 92–93
pore activation index, 13
posterior–anterior, 108

post-prandial, 159, 164
post-processing, 62, 74
prandial, 160
pre-diabetes, 156–158
prepare, 213, 217
pre-processing, 10, 16, 58, 87
pre-screening, 121, 143
pre-trained, 18, 91–93, 176–177
principle component analysis, 11
pristane-induced, 101, 104, 113
pristane induced arthritis, 101, 104
prognosis, 21
prognostic, 172
projections, 7
proximal, 16–17, 48, 124
proximal interphalangeal joints, 16–17
pseudocolor, 55, 57
psychometric, 14

Q

quadratic, 19
quantization, 77
quantum, 155
questionnaire, 122

R

radial, 9
radiographic, 22–23, 106, 122
radiometric, 2–3, 31, 49, 55, 57, 184
random forest, 82, 141–142, 144, 175
Raynaud's, 2, 31, 46–47
recognition, 14–15, 58, 60, 62, 66, 73, 75, 84,
        91, 93, 132, 135, 174, 193, 214
reflectivity, 39
region-based, 68–70
region-growing, 68–69
region-merging, 68
registered, 62, 169
regularization, 87–88
regulation-thermography, 37
reperfusion, 46, 68
resampling, 83
rescaled, 13
ResNet V2, 92
resolutions, 30, 111
respiration-monitoring, 8
respiratory signal, 7, 9
retrospective, 22–23
rewarming, 16, 19, 46
rheumatoid, 1, 16–18, 97, 121, 125, 127, 132,
        145, 149
rheumatoid arthritis, 97, 121, 125, 127, 132,
        145, 149
Robert, 70–71, 83, 132–134

rotated BRIEF feature descriptor, 175
rotational, 61, 138, 177
rotational–invariant, 175

S

sagittal, 20, 44
scale-invariant, 74, 79, 174–175
scale-space, 79
sedentary, 156
segmentation-based, 80
segmented, 7, 13, 17–18, 77, 80, 105, 131,
    135–137, 147–150
semi-automated, 10, 64
semi-quantitative, 103, 146
Shapiro–Wilk, 107
sharpening, 58–59
similarity, 40, 47, 61–62, 75, 81, 84, 138
similarity-based, 144
simplified disease activity index, 124
skeletonization algorithm, 16
smoothening kernel, 58
Sobel, 7, 70–71, 132, 134, 136
Sobel edge detection, 70, 134
Sobel operator, 7, 136
softmax, 92–93, 177
software-assisted, 128
software-based, 128
spatial filtering, 58
spatial–temporal, 7, 11
specific heat capacity, 3
speckle, 46
spectra, 9, 11
spectral, 2–3, 5–6, 60
spectrum, 9, 60, 84, 130, 164
speeded-up, 74–75, 174
speeded-up robust features, 74, 174
stability, 31, 83, 87, 140, 216, 218
stabilization, 37
standardized, 29–30, 46, 112, 160, 186
statistical, 17, 41, 74–75, 83, 107–108,
    138, 163
statistical features, 17
straightforward, 61, 88, 178
subcutaneous, 1, 12, 97
subcutaneous skin temperature, 1
sub-image, 194, 199
subsample, 183
subsampling, 85–86
superficial, 9, 49, 118, 121, 157, 191
superimposed, 136–138, 148–150
support vector machine, 138, 169, 175, 198
SURF, 62, 74–75, 138, 174–175
surrogate, 41
symmetry, 19, 40, 47–48

sympathetic, 21, 219
Synchrotron, 102
synovial hyperplasia, 106, 114–117
system-based, 173

T

technologies, 121, 158, 191
Teledyne, 56
tele-thermographic, 205
temporal, 4, 9–10, 36, 191–192
tenderness, 17
TensorFlow, 177
textures, 68, 78
therapeutic, 41
thermal accuracy, 185, 187
thermal conductance, 3
thermal conductivity, 3
thermal diffusivity, 3
thermal equilibrium, 2, 37
thermal gradients, 21, 46, 48
thermal inertia, 3
thermal insulance, 3
thermal radiation, 1, 2, 29, 130, 161
thermal resistance, 1, 3–4
thermal sensitivity, 185, 187
thermal signature, 12
thermal symmetry, 40, 47–48
thermal video frames, 8
thermistor, 5
thermodynamics, 2, 20, 33
thermo-effector loops, 35
thermogram, 6, 60, 124, 129, 185, 190, 219
thermographic index, 124–125, 143
thermometry, 2, 25, 39, 181–182
thermoneutral zone, 33, 182
thermoregulation, 1, 4, 33, 169, 181
thermoregulatory, 4–5, 33–35, 38, 182
thermoregulatory response, 5, 33–34, 182
threshold-based, 67, 130–132
thresholding, 8, 10, 64, 67–68, 79, 131,
    137, 143
time-consuming, 17, 127, 192, 196
topography, 18, 136
traceability, 31
tracking algorithm, 5–6, 11, 13
trajectories, 10
translation, 15, 79, 198
transmission, 25, 58, 214
transposed, 13
transverse, 146
triangulation, 125
t-test, 98, 163
tumor, 17, 20–21, 70, 136, 184
two-class, 140–141

U

ulceration, 168
ultrasonographic, 145–146
uncontrolled, 1, 38, 166, 177
uncooled, 31, 108, 122, 186
uncorrelated, 11
u-net, 64, 73
uniformity, 69, 76, 78, 216
unsupervised, 66, 88, 135
utilization, 131, 206

V

variance, 11, 75–76, 78, 83, 87, 102, 131, 142, 174, 198
vasoconstriction, 4, 12, 34
vasodilation, 4, 182
vectorized, 175
vertical trajectories, 10
video tracking algorithms, 5
visualization, 111

volumetric, 3
voxel, 105

W

watershed-based, 64
waveform, 7, 9
wavelet-based, 7, 8
wavelet transform, 5, 7, 63
whole-body, 21–22
within-class, 141
Wunderlich, C.A, 181

X

x-direction, 71, 198
X-ray, 97, 102, 145

Z

zero-crossings, 72
zero-padding, 6
zones, 78, 169